ELECTRICAL LICENSING

Exam Guide

SECOND EDITION

By H. Ray Holder

Published by:

www.DEWALT.com/guides

DEWALT Electrical Licensing Exam Guide – 2nd Edition
H. Ray Holder

Vice President, Technology and Trades Professional Business Unit: . . . . Gregory L. Clayton
Product Development Manager: . . . . Robert Person
Executive Marketing Manager: . . . . Taryn Zlatin
Marketing Manager: . . . . Marissa Maiella

ISBN-13: 978-0-9797403-1-2
ISBN-10: 0-9797403-1-2

Delmar
5 Maxwell Drive
Clifton Park, NY 12065-2919
USA

Cengage Learning is a leading provider of customized learning solutions with office locations around the globe, including Singapore, the United Kingdom, Australia, Mexico, Brazil and Japan. Locate your local office at: **international.cengage.com/region**.
Cengage Learning products are represented in Canada by Nelson Education, Ltd.

For your lifelong learning solutions, visit **delmar.cengage.com**.
Visit our corporate website at **cengage.com**.

Notice to the Reader
Information contained in this work has been obtained from sources believed to be reliable. However, neither DEWALT, Cengage Learning nor its authors guarantee the accuracy or completeness of any information published herein, and neither DEWALT, Cengage Learning nor its authors shall be responsible for any errors, omissions, damages, liabilities, or personal injuries arising out of use of this information. This work is published with the understanding that DEWALT, Cengage Learning and its authors are supplying information but are not attempting to render engineering or other professional services. If such services are required, the assistance of an appropriate professional should be sought. The reader is expressly warned to consider and adopt all safety precautions and to avoid all potential hazards. The publisher and DEWALT make no representation or warranties of any kind, nor are any such representations implied with respect to the material set forth here. Neither the publisher nor DEWALT shall be liable for any special, consequential, or exemplary damages resulting, in whole or part, from the readers' use of, or reliance upon, this material.

Printed in Canada

2 3 4 5 XX 10 09

Trade Reference

Blueprint Reading
Construction
Construction Estimating
Construction Safety/OSHA
Datacom
Electric Motor
Electrical Estimating
Electrical – 2008 Code
HVAC Estimating
HVAC/R—Master Edition
Lighting & Maintenance
Plumbing
Plumbing Estimating
Residential Remodeling & Repair
Security, Sound & Video
Spanish/English Construction Dictionary—Illustrated Edition
Wiring Diagrams

Exam and Certification

Building Contractor's Licensing
Electrical Licensing
HVAC Technician Certification
Plumbing Licensing

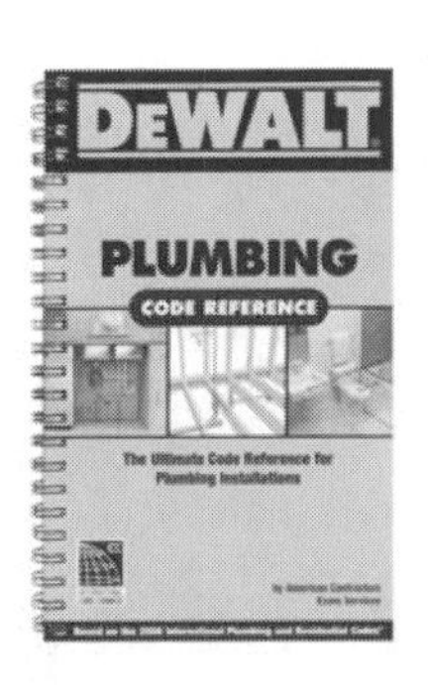

Code Reference

Building
Electrical
HVAC/R
Plumbing

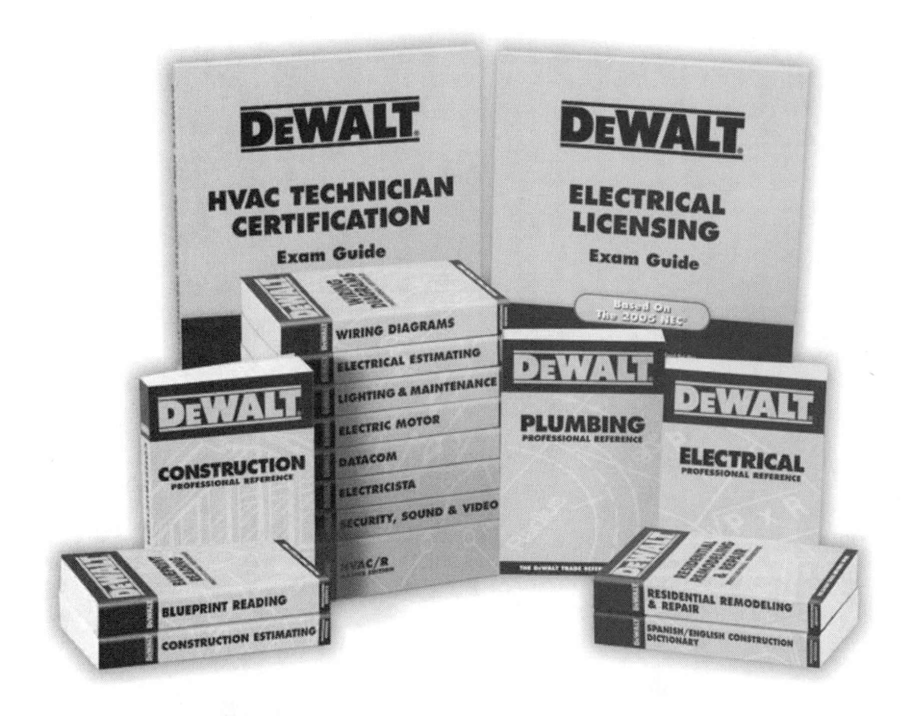

CONTENTS

DeWALT ELECTRICAL LICENSING EXAM GUIDE – SECOND EDITION

by
H. Ray Holder

Notes

HOW TO PREPARE FOR THE EXAM

This book is a guide to preparing for the electricians' exams. It will not make you a competent electrician nor teach you the electrical trade, but it will give you an idea of the type of questions asked on most electricians' exams and how to answer them correctly.

Most exams consist of multiple-choice questions and these are the type reflected in this exam guide. They will give you a feel for how the examinations are structured and are based on many questions the author has encountered while taking numerous exams in recent years.

Begin your pre-exam preparation with two points in mind.

- OPPORTUNITIES in life will ARISE – be prepared for them.
- The more you LEARN – the more you EARN.

Attempting to take an exam without preparation is a complete waste of time. Attend classes at your local community college. Attend seminars, electrical code updates and company-sponsored programs. Many major electrical suppliers and local unions sponsor classes of this type at no cost. Take advantage of them.

Become familiar with the National Electrical Code® (NEC®); the Code has a LANGUAGE all its own. Understanding this language will help you to better interpret the NEC.® Do not become intimidated by its length. Become thoroughly familiar with the definitions in Chapter One; if you don't, the remainder of the NEC® will be difficult to comprehend. Remember, on the job we use different "lingo" and phrases compared to the way the NEC® is written and to the way many test questions are expressed.

HOW TO STUDY

Before beginning to study, get into the right frame of mind and relax. Study in a quiet place that is conducive to learning. If such a place is not available, go to your local library. It is important that you have the right atmosphere in which to study.

It is much better to study many short lengths of time than attempt to study fewer, longer lengths of time. Try to study about an hour every evening. You will need the support and understanding of your family to set aside this much-needed time.

As you study this exam preparation book, the NEC® and other references, always highlight the important points. This makes it easier to locate Code references when taking the exam.

Use a straight edge such as a six-inch ruler when using the NEC® tables and charts. A very common mistake is to get on the wrong line when using these tables; when that happens, the result is an incorrect answer.

Use tabs on the major sections of your NEC® so they are faster and easier to locate when taking the exam. The national average allowed per question is less than three minutes; you cannot waste time.

WHAT TO STUDY

A common reason for one to be unsuccessful when attempting to pass electrical exams is not knowing what to study. Approximately forty percent of most exams are made up of "core" questions. This type of question is reflected in this exam preparation book.

The subject matter covered in most electrical license examinations is:

- Grounding and bonding
- Overcurrent protection
- Wiring methods and installation
- Boxes and fittings
- Box and raceway fill
- Services and equipment
- Motors
- Special occupancies
- Load calculations
- Lighting
- Appliances
- Hazardous locations

Become very familiar with questions on the above. Knowing what to study is a major step toward passing your exam.

HELPFUL HINTS ON TAKING THE EXAM

- **Complete the easy questions first.** On most tests, all questions are valued the same. If you become too frustrated on any one question, it may have a negative effect upon your entire test.
- **Keep track of time.** Do not spend too much time on one question. If a question is difficult for you, mark the answer you think is correct and place a check (✔) by that question in the

Notes

Notes

examination booklet. Then go on to the next question. If you have time after finishing the rest of the exam, you can go back to the questions you have checked off. If you simply do not know the answer to a question, take a guess. Always choose the answer that is most familiar to you.

- **Only change answers if you know you are right.** Usually, your first answer is your best answer.
- **Relax.** Do not get uptight and stressed out when testing.
- **Tab your Code Book.** References are easier and faster to find.
- **Use a straight-edge.** Prevent getting on the wrong line when referring to the tables in the NEC.®
- **Get a good night's rest before the exam.** Do not attempt to drive several hours to an exam site; be rested and alert.
- **Understand the question.** One key word in a question can make a difference in what the question is asking. Underlining key words will help you to understand the meaning of the question.
- **Use a dependable calculator.** Use a solar-powered calculator that has a battery back-up. Because many test sites are not well lighted, this type of calculator will prepare you for such a situation. If possible, bring a spare calculator.
- **Show up at least 30 minutes prior to your exam time.** Be sure to allow yourself time for traffic, etc. when planning your route to the exam location.

REGULATIONS AT THE EXAMINATION SITE

To ensure that all examinees are tested under equally favorable conditions, the following regulations and procedures are observed at most examination sites.

- Each examinee must present proper photo identification, preferably your driver's license, before you will be permitted to take the examination.
- No cameras, notes, tape recorders, pagers or cellular phones are allowed in the examination room.
- No one will be permitted to work beyond the established time limits.
- Examinees are not permitted any reference material EXCEPT the National Electrical Code.®

- Examinees will be permitted to use noiseless calculators during the examination. Calculators which provide programmable ability or pre-programmed calculators are prohibited.
- Permission of an examination proctor must be obtained before leaving the room while the examination is in progress.

TYPICAL EXAMINATION QUESTIONS

The following examples are intended to illustrate typical questions that appear on most electricians' exams.

Example 1

An equipment grounding conductor of a branch circuit shall be identified by which of the following colors?

a. gray
b. white
c. black
d. green

Here you are asked to select from the listed colors the one that is to be used to identify the equipment grounding conductor of a branch circuit. Because Section 250.119 of the NEC® requires that green or green with yellow stripes be the color of insulation used on a grounding conductor (when it is not bare), the answer is (d).

Example 2

A circuit leading to a gasoline dispensing pump must have a disconnecting means _____.

a. only in the grounded conductors.
b. only in the ungrounded conductors.
c. operating independently in all conductors.
d. that simultaneously disconnects both the grounded and ungrounded conductors feeding the dispensing pump.

Here the "question" is in the form of an incomplete statement. Your task is to select the choice that best completes the statement. In this case, you should have selected (d) because Section 514.11 of the NEC® specifies that such a circuit shall be provided with a means to disconnect simultaneously from the source of supply all conductors of a circuit, including the grounded conductor.

Notes

Notes

Example 3

A building or other structure served shall be supplied by only one service EXCEPT one where the capacity requirements are in excess of _____.

a. 800 amperes at a supply voltage of 600 volts or less.
b. 1,000 amperes at a supply voltage of 600 volts or less.
c. 1,500 amperes at a supply voltage of 600 volts or less.
d. 2,000 amperes at a supply voltage of 600 volts or less.

Again, the "question" is in the form of an incomplete statement and your task is to select the choice that best completes the statement. In this case, you are to find an exception. You have to select the condition that has to be met when supplying a building or structure by more than one service. You should have selected (d) because Section 230.2(C)(1) requires the conditions listed in (d), but does not require or permit the conditions listed in (a), (b) or (c).

Example 4

Disregarding exceptions, the MINIMUM size overhead service-drop conductor shall be _____ AWG copper.

a. 6
b. 8
c. 12
d. 14

Here the "question" is in the form of a fill in the blank and your task is to select the choice that best completes the statement. In this case, exceptions are not applicable. You have to select the minimum size conductor required for overhead service-drop conductors. You should have selected (b) because Section 230.23(B) specifies that the conductors shall not be smaller than size 8 AWG copper.

HOW TO USE THIS BOOK

Each practice exam contained in this book consists of 25 questions. The time allotted for each practice exam is 75 minutes or 3 minutes per question. The final exams vary in time, length, and difficulty depending upon the exam level. Using this time limit as a standard, you should be able to complete an actual examination in the allotted time.

To get the most out of this book you should answer every question and highlight your NEC® for future reference. If you have difficulty with a question and cannot come up with an answer that is familiar to you, put a check mark next to the question and come back to it after completing the remainder of the questions. Review your answers with the **ANSWER KEY** located in the back of this book. This will help you identify your strengths and weaknesses. When you discover you are weaker in some areas than others, you will know that further study is necessary in those areas.

Do only one practice exam contained in this book during an allotted study period. This way you do not get burned out and fatigued trying to study for too long a period. This also helps you develop good study habits.

GOOD LUCK!

Notes

HELPFUL REFERENCES

OHMS LAW CIRCLE FORMULAS

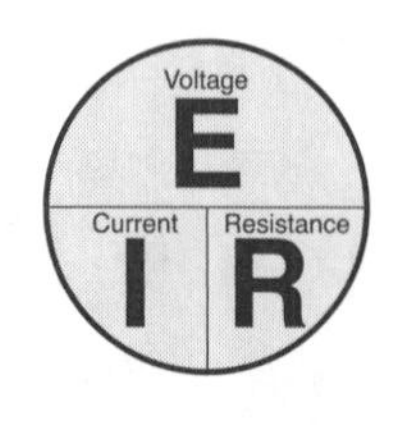

Ohms Law Circle

I = E/R
To Find Current

E = IxR
To Find Voltage

R = E/I
To Find Resistance

POWER CIRCLE FORMULAS

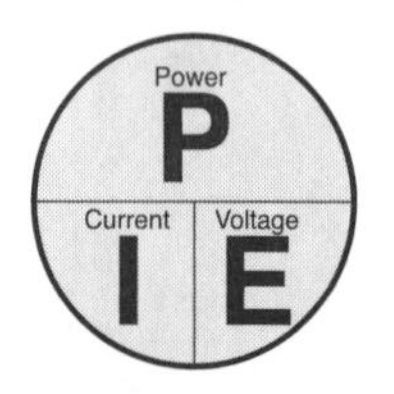

Power Circle

I=P/E
To Find Current

P=xE
To Find Power

E=P/I
To Find Voltage

OHMS LAW/POWER COMBINATION CIRCLE FORMULAS

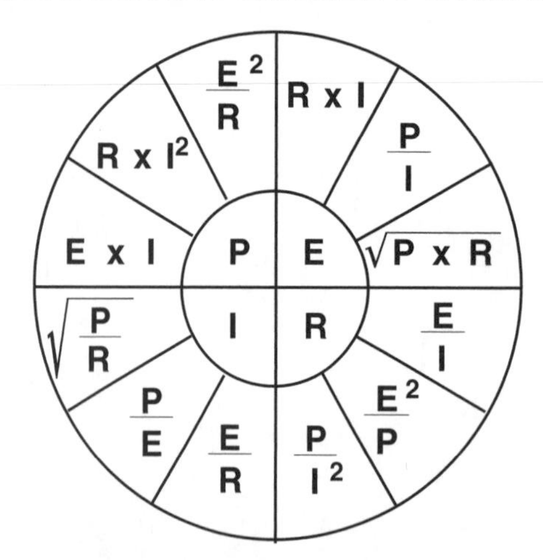

P = Power = Watts or Volt-Amperes
R = Resistance = Ohms
I = Current = Amperes
E = Force = Volts

TYPES OF POWER

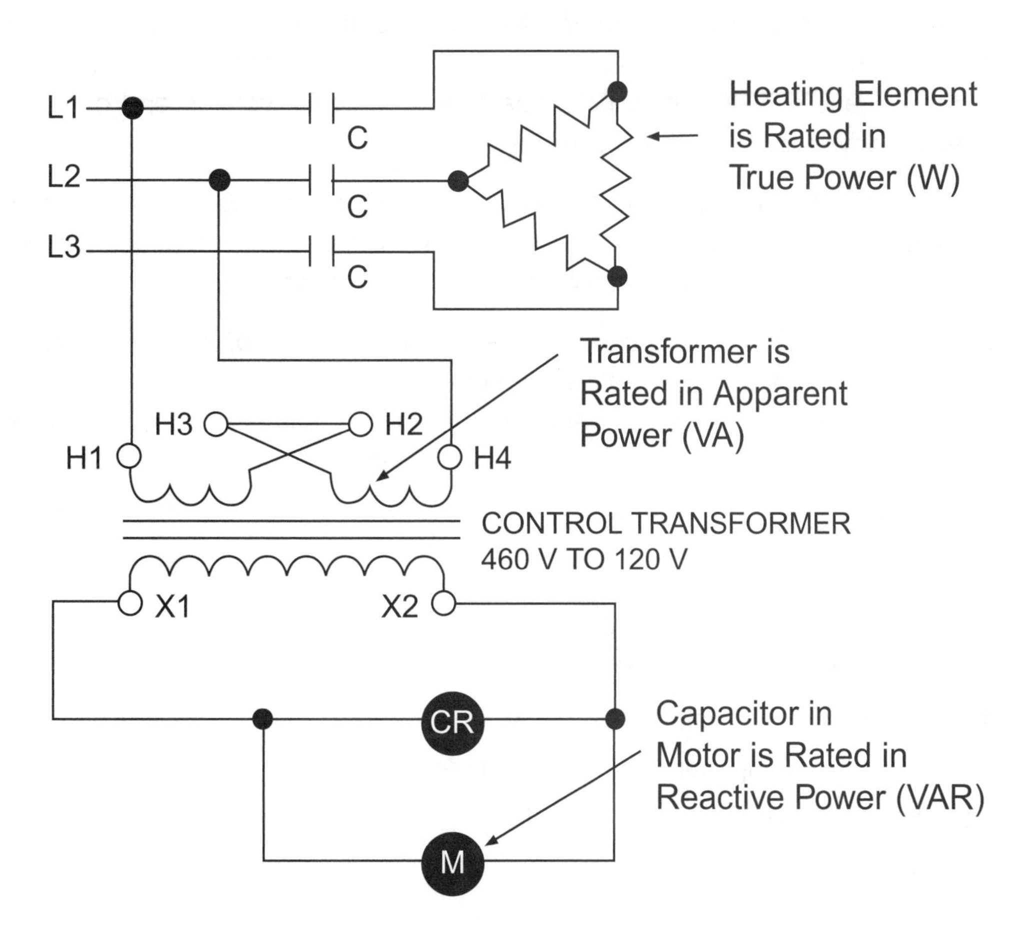

Reactive power is supplied to a reactive load (capacitor/coil) and is measured in volt-amperes reactive (VAR). The capacitor on a motor uses reactive power to keep the capacitor charged. The capacitor uses no true power because it performs no actual work such as producing heat or motion.

In an AC circuit containing only resistance, the power in the circuit is true power. However, almost all AC circuits include capacitive reactance (capacitors) and/or inductive reactance (coils). Inductive reactance is the most common because all motors, transformers, solenoids and coils have inductive reactance.

Apparent power represents a load or circuit that includes both true power and reactive power and is expressed in volt-amperes (VA), kilovolt-amperes (kVA) or megavolt-amperes (MVA). Apparent power is a measure of component or system capacity because apparent power considers circuit current regardless of how it is used. For this reason, transformers are sized in volt-amperes rather than in watts.

TRUE POWER AND APPARENT POWER

True power is the actual power used in an electrical circuit and is expressed in watts (W). Apparent power is the product of voltage and current in a circuit calculated without considering the phase shift that may be present between total voltage and current in the circuit. Apparent power is measured in volt-amperes (VA).

True power equals apparent power in an electrical circuit containing only resistance. True power is less than apparent power in a circuit containing inductance or capacitance. A phase shift exists in most AC circuits that contain devices causing capacitance or inductance.

Capacitance is the property of a device that permits the storage of electrically separated charges when potential differences exist between the conductors. Inductance is the property of a circuit that causes it to oppose a change in current due to energy stored in a magnetic field; i.e., coils.

To calculate true power, apply the formula:

$$P_T = (I)^2 \times R$$

where

P_T = true power (in watts)

I = total circuit current (in amperes)

R = total resistive component of the circuit (in ohms)

To calculate apparent power, apply the formula:

$$P_A = E \times I$$

where

P_A = apparent power (in volt-amperes)

E = measured voltage (in volts)

I = measured current (in amperes)

POWER FACTOR FORMULA

Power factor is the ratio of true power used in an AC circuit to apparent power delivered to the circuit.

$$PF = \frac{P_T}{P_A}$$

where

PF = power factor (percentage)

P_T = true power (in watts)

P_A = apparent power (in volt-amperes)

USEFUL FORMULAS

To Find	Single Phase	Three Phase	Direct Current
Amperes when kVA is known	$\frac{kVA \times 1000}{E}$	$\frac{kVA \times 1000}{E \times 1.732}$	not applicable
Amperes when Horsepower is known	$\frac{HP \times 746}{E \times \%EFF \times PF}$	$\frac{HP \times 746}{E \times 1.732 \times \%EFF \times PF}$	$\frac{HP \times 746}{E \times \%EFF}$
Amperes when Kilowatts are known	$\frac{kW \times 1000}{E \times PF}$	$\frac{kW \times 1000}{E \times 1.732 \times PF}$	$\frac{kW \times 1000}{E}$
Kilowatts	$\frac{I \times E \times PF}{1000}$	$\frac{I \times E \times 1.732 \times PF}{1000}$	$\frac{I \times E}{1000}$
Kilovolt-Amperes	$\frac{I \times E}{1000}$	$\frac{I \times E \times 1.732}{1000}$	not applicable
Horsepower	$\frac{I \times E \times \%EFF \times PF}{746}$	$\frac{I \times E \times 1.732 \times \%EFF \times PF}{746}$	$\frac{I \times E \times \%EFF}{746}$
Watts	$E \times I \times PF$	$E \times I \times 1.732 \times PF$	$E \times I$

I = Amperes
E = Volts
kW = Kilowatts
kVA = Kilovolt-Amperes
HP = Horsepower
%EFF = Percent Efficiency
PF = Power Factor

VOLTAGE DROP FORMULAS

The NEC® recommends a maximum 3% voltage drop for either the branch circuit or the feeder.

Single Phase:

$$VD = \frac{2 \times K \times I \times D}{CM}$$

Three Phase:

$$VD = \frac{1.732 \times K \times I \times D}{CM}$$

VD = Volts (voltage drop of the circuit)

K = 12.9 Ohms/Copper or 21.2 Ohms/Aluminum (resistance constants for a conductor that is 1 foot long and 1 circular mil in diameter at an operating temperature of 75°C.)

I = Amperes (load at 100 percent)

D = Distance in feet (length of circuit from load to power supply)

CM = Circular-Mils (conductor wire size)

2 = Single-Phase Constant

1.732 = Three-Phase Constant

CONDUCTOR LENGTH/VOLTAGE DROP

Voltage drop can be reduced by limiting the length of the conductors.

Single Phase:

$$D = \frac{CM \times VD}{2 \times K \times I}$$

Three Phase:

$$D = \frac{CM \times VD}{1.732 \times K \times I}$$

CONDUCTOR SIZE/VOLTAGE DROP

Increase the size of the conductor to decrease the voltage drop of the circuit (reduce its resistance).

Single Phase:

$$CM = \frac{2 \times K \times I \times D}{VD}$$

Three Phase:

$$CM = \frac{1.732 \times K \times I \times D}{VD}$$

STATE LICENSING REQUIREMENTS FOR ELECTRICAL CONTRACTORS

State	Licensing Board Phone Number	Licensing Board Website	State Licensing Exam/ Testing Company	Pre-licensing or Pre-approval	Continuing Education
Alabama	(334) 272-5030	www.aecb.state.al.us	Yes/PSI	Yes/Pre-approval	No
Alaska	(907) 465-3035	www.dced.state.ak.us/ occ/pcon.htm	Yes/Thomson Prometric	No	Yes
Arizona	(602) 542-1525	www.rc.state.az.us/	Yes/Thomson Prometric	No	No
Arkansas	(501) 682-4548	www.healthyarkansas.com	Yes/Thompson Prometric	Yes/Pre-approval	Yes
California	(800) 321-2752	www.cslb.ca.gov	Yes/California State Licensing Board	Yes/Pre-approval	No
Colorado	(303) 894-2300	www.dora.state.co.us/	Yes/Promissor	Yes/Pre-approval	Yes
Connecticut	(860) 713-6135	www.state.ct.us/dcp	Yes/PSI	Yes/Pre-approval	Yes
Delaware	(302) 744-4504	www.dpr.delaware.gov	Yes/Thomson Prometric	Yes	Yes
Florida	(850) 487-1395	www.state.fl.us/dbpr/pro/ elboard/elec_index.shtml	Yes/Department of Business and Professional Regulation	Yes/Pre-approval	Yes
Georgia	(478) 207-1416	www.sos.state.ga.us/plb/ construct/	Yes/AMP	Yes/Pre-approval	Yes
Hawaii	(808) 586-2689	www.hawaii.gov/dcca/pvl	Yes/Thomson Prometric	Yes/Pre-approval	Yes
Idaho	(208) 334-3233	www2.state.id.us/	Yes/International Code Council	Yes/Pre-approval	Yes
Illinois	Not Regulated by State/contact city/county		No		
Indiana	Not Regulated by State/ contact city/county		No		
Iowa	(515) 281-7995	www.iowaworkforce.org/	No/Register with State; City/County may require testing		
Kansas	Not Regulated by State/ contact city/county		No		
Kentucky	(502) 573-0364	www.ohbc.ky.gov	Yes/International Code Council	No	Yes
Louisiana	(225) 765-2301	www.lslbc.state.la.us	Yes/Louisiana State Licensing Board	Yes/Pre-approval	No
Maine	(207) 624-8603	www.maineprofessionalreg.org	Yes/Electricians Examining Board	Yes/Pre-approval	Yes
Maryland	(410) 230-6163	www.dllr.state.md.us	Yes/PSI	Yes/Pre-approval	No
Massachusetts	(617) 727-9931	www.state.ma.us/reg/ boards/el/default.htm	Yes/Thomson Prometric	Yes/Pre-approval	Yes
Michigan	(517) 241-9320	www.michigan.gov/dleg	Yes/State Electrical Board	No	No
Minnesota	(651) 284-5064	www.electricity.state.mn.us	Yes/State Electrical Board	Yes	Yes

STATE LICENSING REQUIREMENTS FOR ELECTRICAL CONTRACTORS (CONTINUED)					
Mississippi	(601) 354-6161	www.msboc.state.ms.us	Yes/PSI	Yes/Pre-approval	No
Missouri	Not Regulated by State/ contact city/county		No		
Montana	(406) 444-7734	www.electrician.mt.gov	Yes/LaserGrade	Yes/Pre-approval	No
Nebraska	(402) 471-3550	www.electrical.state.ne.us	Yes/State Electrical Board	Yes/Pre-approval	Yes
Nevada	(702) 486-1100	www.nscb.state.nv.us	Yes/PSI	Yes/Pre-approval	No
New Hampshire	(603) 271-3748	www.state.nh.us/electrician	Yes/Bureau of Electrical Safety and Licensing	Yes/Pre-approval	No
New Jersey	(973) 504-6420	www.state.nj.us	Yes/Thomson Prometric	Yes/Pre-approval	Yes
New Mexico	(505) 452-8311	http://www.rld.state.nm.us/ CID/index.htm	Yes/PSI	Yes/Pre-approval	Yes
New York	Not Regulated by State/ contact city/county		No		
North Carolina	(919) 733-9042	www.ncbeec.org	Yes/Promissor	Yes/Pre-approval	Yes
North Dakota	(701) 328-9522	www.ndseb.com	Yes/State Electrical Board	Yes/Pre-approval	Yes
Ohio	(614) 644-3493	www.com.state.oh.us/odoc/dic/ diccont.htm	Yes/International Code Council	Yes/Pre-approval	Yes
Oklahoma	(405) 271-5217	www.cib.state.ok.us/	Yes/PSI	Yes/Pre-approval	Yes
Oregon	(503) 378-4133	www.oregonbcd.org	Yes/PSI	Yes/Pre-approval	Yes
Pennsylvania	Not Regulated by State/contact city/county		No/city/county may require testing		
Rhode Island	(401) 462-8000	www.dlt.state.ri.us	Yes/State Electrical Board	Yes/Pre-approval	Yes
South Carolina	(803) 896-4686	www.llr.state.sc.us	Yes/PSI	No	No
South Dakota	(600) 233-7765	www.state.sd.us/dcr/electrical	Yes/State Electrical Board	Yes/Pre-approval	Yes
Tennessee	(615) 741-8307	www.state.tn.us/commerce/ boards/contractors/	Yes/PSI	No	No
Texas	(512) 463-6599	www.license.state.tx.us/	Yes/International Code Council	No	Yes
Utah	(801) 530-6628	www.dopl.utah.gov	Yes/Thomson Prometric	Yes/Pre-approval	Yes
Virginia	(804) 367-8511	www.state.va.us/dpor	Yes/PSI	Yes/Pre-licensing	No
Washington	(360) 902-5269	www.lni.wa.gov	Yes/LaserGrade	Yes/Pre-approval	Yes
West Virginia	(304) 558-2191	www.wvfiremarshal.org	Yes/Thomson Prometric	Yes/Pre-approval for Master Electrician Only	No
Wisconsin	(608) 261-8500	www.commerce.wi.gov	Yes/State Licensing Board	Yes/Pre-approval	Yes
Wyoming	(307) 777-7288	http://wyofire.state.wy.us/	Yes/International Code Council	Yes/Pre-approval	Yes

MAINTENANCE ELECTRICIAN

Practice Exam #1

The following questions are based on the 2008 edition of the National Electrical Code® and are typical of questions encountered on most Maintenance Electricians' Exams. Select the best answer from the choices given and review your answers with the answer key included in this book.

ALLOTTED TIME: 75 minutes

Notes

Maintenance Electrician Practice Exam #1

1. The National Electrical Code® (NEC®) is NOT:

    a. designed for future expansion of electrical use.
    b. designed to safeguard people and property from electrical hazards.
    c. published by the NFPA®.
    d. intended as a specification manual for trained persons.

2. The NEC® mandates specific branch-circuits, receptacle outlets, and utilization equipment to be provided with a ground-fault circuit interrupter (GFCI); this device is intended _____.

    a. to prevent overloading the conductors
    b. to prevent overloading the circuit breakers
    c. for the protection of equipment from overloads
    d. for the protection of personnel

3. Electrical wiring installed _____ is considered to be installed in a damp location.

    a. under canopies or roofed open porches
    b. underground
    c. outside
    d. none of these apply

4. When a 20-ampere, 120-volt receptacle outlet is installed, in which of the following listed locations is the receptacle required to have GFCI protection?

    a. in a classroom of an educational facility
    b. in the lobby of a movie theater
    c. under an outdoor canopy, near the entrance of a public library
    d. in the walkway of an indoor retail shopping mall

5. Which of the following is a unit of electrical power?

   a. watt
   b. voltage
   c. resistance
   d. conductance

6. When checking for continuity between a circuit breaker and the neutral bar of a panelboard when using a continuity tester, positive continuity is indicated. The reason may be _____.

   I. a conductor grounded
   II. a luminaire or an appliance may be turned on

   a. I only
   b. II only
   c. either I or II
   d. neither I nor II

7. When installing electrical metallic tubing (EMT), there shall be no more than _____ 90° bends in the tubing between boxes and/or pull points.

   a. three
   b. four
   c. five
   d. six

8. The full-load running current of a 3-HP, 208-volt, single phase, continuous duty AC motor is _____.

   a. 10.6 amperes
   b. 19.6 amperes
   c. 13.2 amperes
   d. 18.7 amperes

Notes

Notes

9. When a 225-ampere rated panelboard contains only twenty snap switches (circuit breakers) rated at 30 amperes each, the panelboard shall have overcurrent protection not in excess of _____ amperes.

   a. 200
   b. 225
   c. 300
   d. 600

10. If the voltage-drop of a branch-circuit is too great, which of the following, if any, may be the result?

    I. unsatisfactory illumination of lighting fixtures
    II. overheating of a motor or unsatisfactory motor speed

    a. I only
    b. II only
    c. both I and II
    d. neither I nor II

11. What is the MINIMUM required distance from the floor to the ceiling in front of a 400-ampere rated motor control center, disconnect switch, or main switchboard?

    a. 6 feet
    b. 6 feet, 3 inches
    c. 6 feet, 6 inches
    d. 8 feet

12. In general, the NEC® mandates the largest size insulated solid conductor permitted to be installed into an existing raceway is _____.

    a. 4 AWG
    b. 6 AWG
    c. 8 AWG
    d. 10 AWG

13. A neutral conductor is a conductor that ____.

    a. shall be identified only by white-colored insulation
    b. is intended to carry current under normal conditions
    c. is not intended to carry current under normal conditions
    d. shall be identified by green-colored insulation only

14. Which one of the following listed circuit breakers is NOT a standard ampere rating listed in the NEC®?

    a. 110 ampere
    b. 90 ampere
    c. 75 ampere
    d. 225 ampere

15. A 4 inch × 1½ inch metal octagon box with a flat blank cover may contain a MAXIMUM of _____ size 14 AWG conductors.

    a. six
    b. seven
    c. nine
    d. ten

16. In general, electrical nonmetallic tubing (ENT) shall be securely fastened at intervals not exceeding _____.

    a. 10 feet
    b. 6 feet
    c. 4 feet
    d. 3 feet

Notes

Notes

17. In general, when a 20-ampere rated branch-circuit supplies a single receptacle outlet, the rating of the receptacle must not be less than _____ amperes.

    a. 10
    b. 15
    c. 16
    d. 20

18. A branch-circuit supplying a 7½ HP, 240-volt, single-phase, induction type, continuous-duty, Design B AC motor shall have an ampacity of at least _____ amperes.

    a. 60
    b. 40
    c. 50
    d. 75

19. Where the copper conductors supplying the previous referenced motor will have a temperature rating of 75°C, determine the MINIMUM size as required by the NEC®.

    a. 8 AWG
    b. 6 AWG
    c. 10 AWG
    d. 4 AWG

20. Which of the following fuses, if any, shall be used for replacement only in existing installations and not permitted for use in new installations?

    I. Edison-base fuse
    II. Class H renewable cartridge fuse

    a. I only
    b. II only
    c. both I and II
    d. neither I nor II

21. A 240-volt, single-phase, 10 KW rated commercial dishwasher has a full-load current rating of _____.

    a. 21 amperes
    b. 24 amperes
    c. 42 amperes
    d. 30 amperes

22. The ampacity of a conductor is defined by the NEC® to be the current in amperes, a conductor can carry continuously under the conditions of use without exceeding ______.

    a. its temperature rating
    b. the allowable voltage drop limitations
    c. its melting point
    d. its rated voltage

23. That portion of a wiring system that is between the final overcurrent device protecting the circuit and the outlets is defined as a _____.

    a. branch-circuit
    b. feeder
    c. sub-feeder
    d. service-drop

24. Given: a 120-volt branch-circuit is to serve only fifteen, 150 watt incandescent luminaires. Determine the full-load current in the circuit.

    a. 36.50 amperes
    b. 18.75 amperes
    c. 9.37 amperes
    d. 14.10 amperes

Notes

Notes

25. In general, all switches and circuit breakers used as switches shall be located so that they may be operated from a readily accessible place. They shall be installed so that the center of the grip of the operating handle of the switch or circuit breaker, when at its highest position is NOT more than _____ above the floor or platform.

    a. 5½ feet

    b. 6 feet

    c. 6½ feet

    d. 6 feet, 7 inches

MAINTENANCE ELECTRICIAN

Practice Exam #2

The following questions are based on the 2008 edition of the National Electrical Code® and are typical of questions encountered on most Maintenance Electricians' Exams. Select the best answer from the choices given and review your answers with the answer key included in this book.

ALLOTTED TIME: 75 minutes

Notes

Maintenance Electrician Practice Exam #2

1. In computing the load of luminaires that employ ballast, transformers, or autotransformers, such as fluorescent lighting units, the load must be based on the _____.

    a. total wattage of the lamps
    b. VA of the lamps
    c. length of the lighting fixture
    d. ampere ratings of the complete units

2. A branch-circuit inverse-time circuit breaker rated in amperes shall be permitted as a controller for _____.

    a. continuous-duty motors
    b. intermittent-duty motors
    c. torque motors
    d. all motors

3. Compliance with the provisions of the NEC® will result in an electrical installation that is essentially _____.

    a. free from hazard
    b. a good electrical system
    c. an efficient system
    d. all of these

4. Under normal conditions, an equipment grounding conductor will _____.

    a. improve current flow
    b. not carry current
    c. carry current
    d. reduce circuit resistance

5. Cord-and-plug-connected vending machines located in break rooms of commercial establishments are required to be provided with _____.

   I. a power supply cord not smaller than size 12 AWG with a length not exceeding 4 feet.

   II. GFCI protection

   a. I only
   b. II only
   c. both I and II
   d. neither I nor II

6. A motor with nine leads coming out of it is a _____.

   a. two-phase motor
   b. two-speed motor
   c. dual-voltage motor
   d. multi-speed motor

7. On a four-wire, delta-connected system, the conductor having the higher voltage to ground (high-leg) shall be identified as _____ in color, if the grounded conductor is also present to supply lighting or similar loads.

   a. white
   b. red
   c. green
   d. orange

8. When an electrical trade size $\frac{3}{4}$ inch electrical metallic tubing (EMT) is in excess of 24 inches in length, the conduit is permitted to contain no more than _____ size 10 AWG conductors with THWN insulation.

   a. six
   b. ten
   c. eight
   d. fourteen

Notes

Notes

9. Given: an electrical metallic tubing (EMT) 50 feet in length to be installed will contain a total of nine conductors. Six of these wires are considered to be current-carrying. What is the derating factor, in percent, that must be applied to the ampacity of the current-carrying conductors?

   a. 80 percent
   b. 70 percent
   c. 60 percent
   d. 50 percent

10. The reason the NEC® requires all grounded and ungrounded conductors of a common circuit to be installed together in the same ferrous metal raceway is to reduce _____.

    a. expense
    b. inductive heat
    c. voltage drop
    d. resistance

11. In general, electrical metallic tubing (EMT) shall be securely fastened within ____ of each junction box, panelboard, or other conduit termination.

    a. 3 feet
    b. 6 feet
    c. 8 feet
    d. 10 feet

12. One horsepower is equal to _____ watts.

    a. 746
    b. 1,000
    c. 1,250
    d. 1,500

13. For continuous-duty motors, the motor nameplate current rating is used to determine the size of the _____ required for the motor.

    a. disconnecting means
    b. branch-circuit conductors
    c. motor overload protection
    d. short-circuit protection

14. In general, branch-circuit conductors that supply a single AC motor used in a continuous-duty application shall have an ampacity of NOT less than _____ percent of the motor's full-load running current rating.

    a. 125
    b. 100
    c. 150
    d. 115

15. Unless listed and identified otherwise, terminations and conductors of circuits rated 100 amperes or larger shall be rated at _____.

    a. 60°C
    b. 75°C
    c. 90°C
    d. 120°C

16. When circuit breakers are used to switch 120-volt and 277-volt, high-intensity discharge lighting branch-circuits, the circuit breakers shall be listed and marked _____.

    I. SWD
    II. HID

    a. I only
    b. II only
    c. either I or II
    d. neither I nor II

Notes

Notes

17. An insulated branch circuit conductor with a continuous yellow-colored insulation is considered to be a/an _____.

    a. equipment grounding conductor
    b. grounded conductor
    c. switch-leg
    d. ungrounded conductor

18. When an insulated black conductor is identified by three continuous white stripes along its entire length, it is considered a/an _____.

    a. grounded conductor
    b. ungrounded conductor
    c. equipment grounding conductor
    d. phase of a delta-connected system

19. Given: a 9 kW rated 208-volt, three-phase electric steamer is to be installed in the kitchen of a commercial establishment, the steamer will draw _____ amperes.

    a. 43
    b. 4.3
    c. 25
    d. 75

20. What is the approximate MAXIMUM distance a single-phase, 240-volt, 42-ampere load may be located from a panelboard, given the following related information?

    - copper conductors, K = 12.9
    - size 8 AWG THWN/THHN conductors are used
    - limit voltage drop to 3 percent

    a. 50 feet
    b. 110 feet
    c. 160 feet
    d. 195 feet

21. The NEC® permits electrical trade size 3/8 inch flexible metal conduit (FMC) to be used for tap conductors (fixture whips) to luminaires, provided the length of the flex does not exceed _____.

    a. 4 feet
    b. 6 feet
    c. 8 feet
    d. 10 feet

22. The MINIMUM required width of the working space in front of panelboards, switchboards, disconnects, and motor controllers shall be the width of the equipment or _____ , whichever is greater.

    a. 2 feet
    b. 3 feet
    c. 4 feet
    d. 30 inches

23. In general, branch-circuit conductors serving continuous loads, such as luminaires in an office building or commercial plant, shall have an ampacity of NOT less than _____ percent of the load.

    a. 125
    b. 115
    c. 150
    d. 80

24. A SPD (surge arrester or TVSS) device may be installed on circuits of up to _____ volts.

    a. 150
    b. 300
    c. 600
    d. 1,000

Notes

Notes

25. In general, the overcurrent protection for size 12 AWG copper conductors, regardless of the insulation type, shall not exceed _____ amperes.

    a. 15

    b. 20

    c. 25

    d. 30

RESIDENTIAL ELECTRICIAN

Practice Exam #3

The following questions are based on the 2008 edition of the National Electrical Code® and are typical of questions encountered on most Residential Electricians' Exams. Select the best answer from the choices given and review your answers with the answer key included in this book.

ALLOTTED TIME: 75 minutes

Notes

Residential Electrician Practice Exam #3

1. Ferrous metal enclosures for grounding electrode conductors shall be _____.

    a. electrically continuous
    b. electrically isolated
    c. not permitted
    d. rigid metal conduit (RMC) only

2. Unless listed for the control of other loads, general use dimmer switches shall only be used to control _____.

    a. switched receptacles for cord-connected incandescent luminaires
    b. permanently installed incandescent luminaires
    c. ceiling fans with luminaires
    d. holiday decorative lighting

3. All 125-volt, single-phase, 15- and 20-ampere receptacles installed in residential garages shall be _____.

    I. protected by a listed arc-fault interrupter
    II. provided with ground-fault circuit-interrupter protection

    a. I only
    b. II only
    c. both I and II
    d. neither I nor II

4. The MAXIMUM allowable ampacity of size 18 AWG fixture wire is _____ amperes.

    a. 12
    b. 10
    c. 8
    d. 6

5. A receptacle outlet installed for a gas or electric clothes dryer in the laundry room of a dwelling unit must be installed within at LEAST _____ feet of the intended location of the dryer.

    a. 6
    b. 4
    c. 10
    d. 3

6. Which of the following listed conductors are required to have overcurrent protection on a residential electric service?

    a. grounding conductors
    b. bonding conductors
    c. identified conductors
    d. ungrounded conductors

7. Given: a 40-gallon electric water heater has a nameplate rating of 4,500 watts @ 240 volts. What is the MAXIMUM standard size circuit breaker the NEC® permits to protect the water heater?

    a. 20 amperes
    b. 25 amperes
    c. 30 amperes
    d. 45 amperes

8. If a cable or raceway enters an enclosure such as a meter base, above uninsulated live parts in a wet location, the fittings used to connect the enclosure must be _____ for wet locations.

    a. identified
    b. marked
    c. suitable
    d. listed

Notes

Notes

9. In general, the power supply cord to a mobile home shall have a MAXIMUM rating of _____.

    a. 50 amperes
    b. 60 amperes
    c. 100 amperes
    d. 125 amperes

10. A 150 kVA, single-phase transformer having a secondary voltage of 120/240 is to be installed at a multi-family dwelling. The current available at the secondary is _____ amperes.

    a. 329
    b. 421
    c. 625
    d. 729

11. When the heating, air-conditioning, or refrigeration equipment is installed on the roof of an apartment complex, a 125-volt, 15- or 20-ampere rated receptacle outlet _____.

    a. is not required by the NEC®
    b. may be connected to the line side of the equipment disconnecting means, if the outlet is GFCI protected
    c. shall be located on the same level and within 25 feet of the heating, air-conditioning, or refrigeration equipment
    d. shall be installed on the roof where the equipment is located and not more than 75 feet from each unit

12. The MINIMUM number of 120-volt, 15-ampere, general lighting branch circuits required for a dwelling that has 70 feet × 30 feet of habitable space is _____.

    a. two
    b. three
    c. four
    d. five

Notes

13. Receptacle outlets on a 20-ampere rated small appliance branch-circuit serving kitchen countertops in a dwelling unit are to be rated _____.

    I. 15 amperes

    II. 20 amperes

    a. I only
    b. II only
    c. either I or II
    d. neither I nor II

14. In general, service conductors installed as open conductors shall have a clearance of not less than _____ from windows that are designed to be opened, doors, porches or balconies.

    a. 6 feet
    b. 8 feet
    c. 5 feet
    d. 3 feet

15. Given: a 120-volt lighting branch-circuit is to supply only six 100 watt, 120-volt, incandescent luminaires. Determine the total current on this branch-circuit.

    a. 30 amperes
    b. 20 amperes
    c. 5 amperes
    d. 12 amperes

16. Raceways installed in solid rock shall be permitted to be buried at lesser depths than generally required, when covered by at least _____ of concrete, extending down to the rock.

    a. 2 inches
    b. 4 inches
    c. 6 inches
    d. 8 inches

Notes

17. Electrical outlet boxes installed in walls and ceilings are required to be _____.

    a. visible
    b. readily accessible
    c. accessible
    d. metal

18. The MINIMUM internal depth of a box that is intended to enclose a flush mounted device shall be _____.

    a. $\frac{1}{2}$ inch
    b. $\frac{9}{16}$ inch
    c. $\frac{15}{16}$ inch
    d. $1\frac{1}{4}$ inches

19. Metallic surface type cabinets for electrical equipment located in damp or wet locations shall be mounted so there is at LEAST _____ air space between the cabinet and the wall or other supporting surface.

    a. $\frac{1}{8}$ inch
    b. $\frac{1}{4}$ inch
    c. $\frac{3}{8}$ inch
    d. $\frac{1}{2}$ inch

20. In dwelling units, hallways of _____ feet or more in length are required to have at least one-, 15- or 20-ampere rated 120-volt receptacle outlet.

    a. 10
    b. 12
    c. 15
    d. 20

21. Receptacles in a kitchen of a residence that are to serve countertop surfaces shall be installed so that no point along the wall line is more than _____ inches, measured horizontally from a receptacle outlet in that space.

    a. 24
    b. 18
    c. 36
    d. 48

22. Given: you are to install a 20-ampere, 120-volt, GFCI protected, buried, UF cable branch-circuit that supplies landscape lighting for a residence. The UF cable is required to have a ground cover of at LEAST _____.

    a. 6 inches
    b. 12 inches
    c. 18 inches
    d. 24 inches

23. Where nonmetallic sheathed cable (NM) is run at angles with joists in unfinished basements, the smallest three conductor cable permitted to be installed directly to the lower edges of the joists is size _____.

    a. 12 AWG
    b. 10 AWG
    c. 8 AWG
    d. 6 AWG

24. When exceptions are not to be taken into consideration, the ungrounded conductor of an electrical branch-circuit may be identified by _____.

    a. white- or gray-colored insulation
    b. bare or green-colored insulation
    c. only gray-colored insulation
    d. black- or red-colored insulation

Notes

Notes

25. Three-way and four-way switches shall be so wired that all switching is done_____.

    a. only in the grounded circuit conductor
    b. only in the ungrounded circuit conductor
    c. either in the grounded or ungrounded circuit conductor
    d. only in the white colored circuit conductor

RESIDENTIAL ELECTRICIAN

Practice Exam #4

The following questions are based on the 2008 edition of the National Electrical Code® and are typical of questions encountered on most Residential Electricians' Exams. Select the best answer from the choices given and review your answers with the answer key included in this book.

ALLOTTED TIME: 75 minutes

Notes

Residential Electrician Practice Exam #4

1. The MAXIMUM height that kitchen countertop receptacle outlets may be installed above the countertop shall be _____.

   a. 12 inches
   b. 18 inches
   c. 20 inches
   d. 24 inches

2. For dwelling units, panelboards are permitted to be located in all the areas listed EXCEPT _____.

   a. hallways
   b. garages
   c. kitchens
   d. bathrooms

3. The MAXIMUM number of grounded conductors allowed to be terminated in an individual terminal, if conductors are not paralled, in the neutral bus of a panelboard is _____.

   a. one
   b. two
   c. three
   d. four

4. Flexible cord is considered as protected by a 20 ampere branch circuit breaker if it is _____ and approved for use with a specific listed appliance.

   a. not less than 6 feet in length
   b. size 20 AWG or larger
   c. size 18 AWG or larger
   d. size 16 AWG or larger

5. A 120-volt receptacle is installed on a dwelling unit kitchen island countertop that is 8 feet from the kitchen sink. Which one of the following statements, if any, is correct?

   a. GFCI protection is not required because the receptacle is not within 6 feet of the sink.
   b. GFCI protection is required for all countertop kitchen receptacles.
   c. GFCI protection is not required on receptacles installed on kitchen islands.
   d. None of the above is correct.

6. Where a mobile home has the main service disconnecting means installed outdoors, the disconnecting means shall be installed, so the bottom of the enclosure is NOT less than _____ above finished grade.

   a. 1 foot
   b. 2 feet
   c. 3 feet
   d. 4 feet

7. A GFCI protected receptacle outlet that provides power to a permanently installed swimming pool recirculating pump motor, shall be permitted not closer than _____ feet from the inside wall of the pool.

   a. 6
   b. 5
   c. 12
   d. 15

8. The grounding contacts of branch-circuit receptacle outlets shall be grounded by connection to the _____ conductor.

   a. bonding
   b. neutral
   c. grounded
   d. equipment grounding

Notes

Notes

9. Any one cord-and-plug-connected utilization equipment connected to a 120-volt, 20-ampere rated branch-circuit shall have a MAXIMUM rating of _____.

   a. 10 amperes
   b. 16 amperes
   c. 20 amperes
   d. 25 amperes

10. In walls or ceilings with a surface of concrete, tile, gypsum, plaster, or other noncombustible material, boxes employing a flush-type cover or faceplate shall be installed so that the front edge of the box will not set back of the finished surface more than _____.

   a. 1/4 inch
   b. 1/2 inch
   c. 3/8 inch
   d. 3/4 inch

11. For a one-family dwelling, the service disconnecting means shall have a rating of not less than _____ amperes when supplied with a 120/240-volt, single-phase service drop.

   a. 30
   b. 60
   c. 100
   d. 200

12. When applying the general method of calculation for a dwelling unit, what is the feeder and service demand load, in kW, for one 8 KW rated residential electric range?

   a. 8.0 kW
   b. 6.4 kW
   c. 7.0 kW
   d. 12.0 kW

13. Service equipment for a mobile home shall be rated not less than _____ at 120/240 volts, single-phase.

    a. 50 amperes
    b. 60 amperes
    c. 100 amperes
    d. 150 amperes

14. Underwater luminaires installed in swimming pools are required to have GFCI protection if they operate at a voltage greater than _____.

    a. 15 volts
    b. 50 volts
    c. 120 volts
    d. 150 volts

15. Given: the ungrounded service-entrance conductors for a one-family residence are size 3/0 AWG THWN copper conductors. A copper grounding electrode conductor attached to the concrete-encased steel reinforcing bars used as the grounding electrode shall not be smaller than size _____.

    a. 2 AWG
    b. 4 AWG
    c. 6 AWG
    d. 8 AWG

16. What is the MAXIMUM period allowed for temporary outdoor Christmas decoration lighting for residences?

    a. 30 days
    b. 60 days
    c. 90 days
    d. 4 months

Notes

Notes

17. When doing box fill calculations, a fixture stud in a ceiling mounted outlet box is an equivalent to _____ conductor(s).

    a. zero
    b. one
    c. two
    d. three

18. A branch-circuit supplying more than one electric baseboard heater in a residential occupancy shall be rated a MAXIMUM of _____.

    a. 15 amperes
    b. 20 amperes
    c. 30 amperes
    d. 50 amperes

19. Disregarding exceptions, when installing electrical metallic tubing (EMT), the run of tubing is required to be securely fastened at LEAST every _____.

    a. 6 feet
    b. 10 feet
    c. 15 feet
    d. 20 feet

20. When calculating the demand load for a one-family dwelling, what total MINIMUM volt-amperes (VA) must be included in the calculation for the small-appliance and laundry circuit loads?

    a. 4,500 VA
    b. 3,000 VA
    c. 1,500 VA
    d. 6,000 VA

21. The NEC® mandates the ampacity of UF cable shall be that of _____ conductors.

    a. 60°C rated
    b. 75°C rated
    c. 86°F rated
    d. 90°C rated

22. The front edge of a switch box installed in a wall constructed of wood, shall be _____ from the surface of the wall.

    a. flush with or projected out
    b. set back a maximum of ¼ inch
    c. set back a maximum of ½ inch
    d. set back a maximum of ⅜ inch

23. For dwelling units, where 125 volt, single-phase, 15 and 20 ampere receptacle outlets are installed within at LEAST _____ of the outside edge of a sink in a laundry room, GFCI protection must be provided.

    a. 4 feet
    b. 6 feet
    c. 8 feet
    d. 10 feet

24. For dwelling units, a branch-circuit supplying _____ receptacle outlets is permitted to also supply receptacle outlets in an attached garage.

    a. outdoor
    b. bathroom
    c. laundry room
    d. kitchen small-appliance

Notes

Notes

25. A metal underground water pipe may serve as a grounding electrode if it is in direct contact with the earth for at LEAST _____ feet or more.

    a. 6
    b. 8
    c. 10
    d. 12

RESIDENTIAL ELECTRICIAN
Practice Exam #5

The following questions are based on the 2008 edition of the National Electrical Code® and are typical of questions on most Residential Electricians' Exams. Select the best answer from the choices given and review your answers with the answer key included in this book.

ALLOTTED TIME: 75 minutes

Notes

Residential Electrician Practice Exam #5

1. What is the MAXIMUM allowable cord length for the cord supplying a cord-and-plug connected dishwasher installed under a kitchen counter in a dwelling unit?

    a. 1½ feet
    b. 2 feet
    c. 3 feet
    d. 4 feet

2. What is the MINIMUM size copper SE cable with type THHW insulation that may be used as ungrounded, service-entrance phase conductors for a 150 ampere-rated, 120/240-volt, single-phase, residential service?

    a. 1/0 AWG
    b. 1 AWG
    c. 2 AWG
    d. 3 AWG

3. For outlet boxes designed to support ceiling-suspended (paddle) fans that weigh more than _____ pounds, the required marking shall include the maximum weight to be supported.

    a. 20
    b. 30
    c. 35
    d. 70

4. A connection to a driven or buried grounding electrode shall ______.

    a. be accessible
    b. not be required to be accessible
    c. not permitted to be buried
    d. be visible

5. What is the MAXIMUM allowable standard size circuit breaker rating that may be used to protect a size 8 AWG copper NM-B cable supplying an electric range?

   a. 30 amperes
   b. 40 amperes
   c. 50 amperes
   d. 60 amperes

6. The MAXIMUM number of size 10 AWG conductors permitted in a 4 inch × 1¼ inch octagon metal junction box is _____.

   a. two
   b. four
   c. five
   d. six

7. Determine the MINIMUM number of 15-ampere, 120-volt general lighting branch-circuits required for a dwelling that has 2,600 square feet of habitable space.

   a. three
   b. four
   c. five
   d. six

8. When calculating the total load on a dwelling unit, how many VA per square feet must be included for the general purpose receptacle outlets?

   a. none
   b. one
   c. two
   d. three

Notes

Notes

9. When calculating the total load for a dwelling unit, what is the MINIMUM load, in VA, that must be added for the two required small appliance branch-circuits?

    a. 1,200 VA
    b. 1,500 VA
    c. 2,400 VA
    d. 3,000 VA

10. It shall be permissible to apply a demand factor of _____ percent to the nameplate-rating load of four or more fastened in place electric water heaters in a multi-family dwelling unit.

    a. 50
    b. 75
    c. 80
    d. 90

11. In dwelling units, at least one wall receptacle shall be installed in bathrooms; the receptacle outlet shall be within at LEAST _____ of the outside edge of each basin.

    a. 12 inches
    b. 18 inches
    c. 24 inches
    d. 36 inches

12. When installing NM cable through bored holes in wooden studs, the holes shall be bored so that the edge of the hole is not less than _____ from the edge, or the cable shall be protected by a steel plate at least 1⁄16 inch thick.

    a. 3⁄4 inch
    b. 1 inch
    c. 1 1⁄4 inches
    d. 1 1⁄2 inches

13. A 240-volt, single-phase, 30-ampere rated, electric clothes dryer has a VA rating of _____.

    a. 5,000 VA
    b. 7,200 VA
    c. 8,000 VA
    d. 15,400 VA

14. Type XHHW insulated conductors are permitted to be used in _____.

    a. dry locations only
    b. wet locations only
    c. dry or damp locations only
    d. dry, damp, or wet locations

15. When two single-pole switches are mounted on the same strap, the number of conductors permitted in the box shall be reduced by _____ conductor(s).

    a. one
    b. two
    c. three
    d. four

16. When doing residential service and feeder calculations, electric clothes dryers are to be calculated at a MINIMUM of _____ watts (VA) or the nameplate rating, whichever is larger.

    a. 3,000
    b. 4,500
    c. 5,000
    d. 6,000

Notes

Notes

17. The rating of any single cord-and-plug connected appliance supplied by a 30-ampere rated branch-circuit shall not exceed _____.

    a. 30 amperes
    b. 27 amperes
    c. 24 amperes
    d. 16 amperes

18. In general, all receptacles installed on 15- and 20-ampere, 120-volt branch-circuits are required to _____.

    a. be GFCI protected
    b. have a nonmetallic faceplate
    c. be of the ungrounded type
    d. be of the grounding type

19. A branch-circuit supplying a 5 kW wall-mounted oven and a 7 kW counter-mounted cooktop in a residence will have a demand load of _____ on the ungrounded service-entrance conductors when applying the general (standard) method of calculation for dwellings.

    a. 12 kW
    b. 9.5 kW
    c. 8.0 kW
    d. 7.8 kW

20. The MINIMUM size copper equipment grounding conductor required to equipment served by a 40-ampere rated branch-circuit is _____.

    a. 10 AWG
    b. 8 AWG
    c. 12 AWG
    d. 14 AWG

21. What is the MAXIMUM allowable voltage between conductors on a branch-circuit supplying luminaires in a residence?

    a. 120 volts
    b. 150 volts
    c. 240 volts
    d. 250 volts

22. The ampacity of the branch-circuit conductors to a residential central heating electric furnace shall NOT be less than _____ percent of the furnace load.

    a. 80
    b. 100
    c. 115
    d. 125

23. Switches and circuit breakers used as a disconnecting means to appliances shall be _____.

    a. factory installed only
    b. of the fusible type
    c. of the indicating type
    d. of the tripping type

24. The NEC® requires recessed portions of luminaire enclosures that are not identified for contact with insulation, to be spaced from combustible material a MINIMUM of _____.

    a. 3/8 inch
    b. 1/2 inch
    c. 3/4 inch
    d. 1 inch

Notes

Notes

25. When installing luminaires without GFCI protection above an indoor spa or hot tub, they are required to be mounted at LEAST _____ above the maximum water level of the spa or hot tub.

    a. 5 feet
    b. 7 feet, 6 inches
    c. 6 feet, 7 inches
    d. 12 feet

RESIDENTIAL ELECTRICIAN
Practice Exam #6

The following questions are based on the 2008 edition of the National Electrical Code® and are typical of questions on most Residential Electricians' Exams. Select the best answer from the choices given and review your answers with the answer key included in this book.

ALLOTTED TIME: 75 minutes

Notes

Residential Electrician Practice Exam #6

1. When a single branch-circuit with an equipment grounding conductor is supplying a detached garage from a house, what, if any, additional grounding electrode(s) is/are required at the garage?

   a. A ground rod and underground metal water pipe.
   b. A metal underground water pipe only.
   c. A metal underground water pipe and building steel, if available.
   d. No additional electrodes are required.

2. The point of attachment of a service drop to a residence where the voltage is 120 volts to ground is required to be at LEAST _____ above grade.

   a. 8 feet
   b. 10 feet
   c. 12 feet
   d. 15 feet

3. The branch-circuit overcurrent protection device shall be permitted to serve as the disconnecting means for stationary motors rated NOT greater than _____ horsepower.

   a. 2
   b. 1
   c. $1/2$
   d. $1/8$

4. When installing an overhead service using a rigid metal conduit (RMC) mast for the support of the service-drop conductors, the mast shall be of adequate strength or be _____ to withstand safely the strain imposed by the service drop.

   a. at least 2 inches in diameter
   b. a minimum of 3 inches in diameter
   c. supported by braces or guys
   d. less than 4 feet in length

5. When wall switches are installed in the same room with a spa or hot tub, the MINIMUM distance from the switches to the inside wall of the hot tub or spa shall be _____.

   a. 5 feet
   b. 10 feet
   c. 15 feet
   d. 18 feet

6. All 15- and 20-ampere, single-phase, 125-volt receptacle outlets located at LEAST _____ from the inside walls of a swimming pool shall be provided with GFCI protection.

   a. 10 feet
   b. 15 feet
   c. 20 feet
   d. 25 feet

7. A luminaire shall be supported independently of the outlet box when it weighs more than _____ pounds, unless the outlet box is listed for the weight to be supported.

   a. 6
   b. 25
   c. 35
   d. 50

8. Of the following listed connectors, which type is prohibited for connection of a grounding conductor to enclosures?

   a. sheet metal screws
   b. pressure connectors
   c. clamps
   d. lugs

Notes

Notes

9. A ground rod is required to be driven a MINIMUM depth of _____ feet into the soil.

   a. 4
   b. 6
   c. 8
   d. 10

10. Receptacle outlets are permitted to be installed directly over a bathtub or shower stall _____.

    a. if the outlet has a weatherproof cover
    b. if the outlet is GFCI protected
    c. if the outlet has a weatherproof cover and is GFCI protected
    d. never

11. Receptacle outlets installed in the _____ position in countertops are prohibited.

    a. vertical
    b. horizontal
    c. face-up
    d. all of these are acceptable

12. Receptacle outlets in a kitchen of a residence that are to serve countertop surfaces shall be installed so that no point along the wall line is more than _____ when measured horizontal from a receptacle in that space.

    a. 24 inches
    b. 18 inches
    c. 36 inches
    d. 48 inches

13. The ampacity of branch-circuit conductors and the rating or setting of overcurrent protective devices supplying fixed outdoor electric deicing equipment, shall NOT be less than _____ percent of the full load current of the equipment.

    a. 100
    b. 125
    c. 80
    d. 150

14. When an intermediate metal conduit (IMC) contains conductors of 600 volts or less and is buried under a gravel driveway of a one-family dwelling, the IMC must be buried at LEAST _____.

    a. 12 inches
    b. 18 inches
    c. 24 inches
    d. 30 inches

15. Disregarding exceptions, where residential lighting outlets are installed in interior stairways, there shall be a wall switch provided _____.

    a. near the stairs
    b. every seven steps
    c. at the top and bottom of the stairs if there are more than six steps
    d. at any convenient location

16. Where it is impractical to locate the service head or gooseneck above the point of attachment of the service drop conductors, the service head or gooseneck shall be permitted not further than _____ from the point of attachment.

    a. 3 feet
    b. 4 feet
    c. 6 feet
    d. 2 feet

Notes

Notes

17. A metal box or terminal fitting having separately bushed holes for each conductor shall be used whenever change is made from a conduit to _____.

    a. knob-and-tube wiring
    b. non-metallic sheathed cable
    c. type AC cable
    d. type MC cable

18. Electric water heaters for swimming pools shall have the heating elements subdivided into loads NOT exceeding _____.

    a. 32 amperes
    b. 36 amperes
    c. 48 amperes
    d. 60 amperes

19. Concealed knob and tube wiring shall be permitted to be used only for _____.

    a. dwellings units
    b. extensions of existing installations
    c. accessible installations
    d. installing temporary wiring

20. In dwelling units, all 125 volt, single-phase, 15 and 20 ampere receptacle outlets installed within 6 feet of the outside edge of a sink in a laundry room shall _____.

    a. have a weatherproof cover
    b. have AFCI protection
    c. be a single receptacle
    d. be provided with GFCI protection

21. The largest size copper conductor permitted for nonmetallic sheathed cable (NM) is size _____ AWG.

    a. 4
    b. 2
    c. 6
    d. 1

22. For the purpose of determining box fill, size 12 AWG copper conductors with THHN insulation are to be calculated at _____ per conductor in the box.

    a. 1.75 cubic inches
    b. 2.00 cubic inches
    c. 2.25 cubic inches
    d. 2.50 cubic inches

23. What is the MAXIMUM distance allowed between supports when installing type NM cable?

    a. 3 feet
    b. 4½ feet
    c. 6 feet
    d. 10 feet

24. Disregarding exceptions, the MINIMUM size underground service lateral conductors permitted by the NEC® is _____.

    a. 8 AWG copper
    b. 6 AWG copper
    c. 4 AWG aluminum
    d. 2 AWG aluminum

Notes

Notes

25. How many size 12/2 AWG with ground, type NM cables are permitted to be installed in an outlet box with a total volume of 18 cubic inches when the box contains a duplex receptacle outlet?

    a. one
    b. two
    c. three
    d. four

JOURNEYMAN ELECTRICIAN
Practice Exam #7

The following questions are based on the 2008 edition of the National Electrical Code® and are typical of questions encountered on most Journeyman Electricians' Exams. Select the best answer from the choices given and review your answers with the answer key included in this book.

ALLOTTED TIME: 75 minutes

Notes

Journeyman Electrician Practice Exam #7

1. A listed instantaneous water heater shall be permitted to be subdivided into circuits protected at NOT more than _____.

   a. 60 amperes
   b. 100 amperes
   c. 120 amperes
   d. 150 amperes

2. The interior of enclosures or raceways installed underground shall be considered to be a _____.

   a. wet location
   b. damp location
   c. moist location
   d. dry location

3. The continuity of the grounding conductor system for portable electrical carnival equipment shall be verified ____.

   a. and recorded on an annual basis
   b. and recorded on a quarterly basis
   c. and recorded on a monthly basis
   d. each time the equipment is connected

4. A plate type grounding electrode shall be installed below the surface of the earth NOT less than _____.

   a. 24 inches
   b. 30 inches
   c. 36 inches
   d. 48 inches

5. Determine the MAXIMUM allowable ampacity of a size 1/0 AWG THW copper current-carrying conductor installed in a common raceway with three other current-carrying conductors of the same size. Ambilent temperature is 86°F.

   a. 150 amperes
   b. 105 amperes
   c. 112 amperes
   d. 120 amperes

6. The NEC® defines a continuous-load to be a load where the maximum current on the circuit is expected to continue for at least _____ hours or more.

   a. 3
   b. 4
   c. 6
   d. 8

7. Determine the MINIMUM size THW copper branch-circuit conductors required by the NEC® to supply a three-phase, continuous duty, AC motor that has a full-load current of 70 amperes. Assume all terminations to be rated at 75°C.

   a. 1 AWG
   b. 2 AWG
   c. 3 AWG
   d. 4 AWG

8. Given: you have determined a conductor has a computed allowable ampacity of 75 amperes. What is the MAXIMUM standard ampere rating of the overcurrent protection device the NEC® permits to protect this circuit? This is not a motor circuit or part of a multi-outlet branch-circuit supplying receptacles.

   a. 70 amperes
   b. 75 amperes
   c. 80 amperes
   d. 85 amperes

Notes

Notes

9. A 208-volt, three-phase, 50 HP, squirrel-cage, continuous duty, Design C, AC motor has a full-load running current of _____.

   a. 130 amperes
   b. 143 amperes
   c. 162 amperes
   d. 195 amperes

10. Which of the following metal piping systems contained in a building is/are required to be bonded?

    a. A gas pipe for a gas heater.
    b. A bathroom copper water pipe, connected to an electric water heater.
    c. An air line for compressed air.
    d. All of these need to be bonded.

11. Determine the MINIMUM trade size electrical metallic tubing (EMT) required to enclose eight size 6 AWG copper conductors with THHW insulation, installed in a 60 foot conduit run.

    a. 1 inch
    b. 1¼ inch
    c. 1½ inch
    d. 2 inch

12. When a feeder supplies a continuous-load of 240 amperes, in general, the overcurrent protection device protecting this circuit shall have a MINIMUM rating of _____.

    a. 240 amperes
    b. 300 amperes
    c. 250 amperes
    d. 275 amperes

13. In general, liquidtight flexible metal conduit (LFMC) shall be securely fastened within _____ of each box or other conduit termination.

    a. 4½ feet
    b. 3 feet
    c. 12 inches
    d. 18 inches

14. When a pull box contains conductors of size 4 AWG and larger and a straight pull of the conductors is to be made, the length of the box shall NOT be less than _____ times the trade diameter of the largest conduit entering the box.

    a. six
    b. four
    c. eight
    d. twelve

15. In general, all of the following wiring methods listed are permitted to be used as a feeder for temporary wiring on a construction site EXCEPT _____.

    a. NM cable
    b. NMC cable
    c. type PD cord
    d. type SJO cord

16. Determine the MAXIMUM number of size 14 AWG THHN conductors permitted to be installed in a trade size ⅜ inch flexible metal conduit (FMC) that contains a bare size 14 AWG grounding conductor when the flex has external fittings.

    a. two
    b. three
    c. four
    d. five

Notes

Notes

17. Which of the following listed conduits does the NEC® permit to enclose conductors feeding wet-niche underwater pool lights?

    a. electrical metallic tubing (EMT)
    b. electrical nonmetallic tubing (ENT)
    c. galvanized rigid metal conduit (RMC)
    d. schedule 40 rigid polyvinyl chloride conduit (PVC)

18. Underground wiring is not permitted under a swimming pool or within _____ of the pool unless the wiring is supplying pool equipment.

    a. 10 feet
    b. 6 feet
    c. 5 feet
    d. 8 feet

19. Which of the following listed types of batteries is NOT permitted for use as a source of power for emergency systems?

    a. automotive
    b. lead acid type
    c. alkali type
    d. all of these

20. The MINIMUM rating for a service disconnecting means for a building, other than a one-family dwelling, consisting of three, two-wire circuits shall be _____.

    a. 15 amperes
    b. 30 amperes
    c. 60 amperes
    d. 100 amperes

21. In guest rooms or guest suites of hotels and motels, and sleeping rooms in dormitories, at LEAST _____ general purpose receptacle(s) installed in the room is/are required to be readily accessible.

    a. one
    b. two
    c. three
    d. all

22. When branch-circuit conductors are contained inside a ballast compartment and they are within 3 inches of the ballast, the conductors shall have a temperature rating of at LEAST _____.

    a. 105°C
    b. 90°C
    c. 75°C
    d. 60°C

23. In general, when a size 4 AWG or larger insulated conductor enters a panelboard, cabinet, or enclosure, which of the following must be provided?

    a. a bonding jumper
    b. a grounding clip
    c. an insulated bushing
    d. an insulated grounding conductor

24. Class _____ locations are those that are hazardous because of the presence of easily ignitible fibers or materials producing combustible flyings.

    a. I
    b. II
    c. III
    d. IV

Notes

Notes

25. A single-phase, 240-volt, 15 kVA generator may have a current load of not more than _____ amperes per line.

    a. 62.50

    b. 31.25

    c. 41.66

    d. 52.50

JOURNEYMAN ELECTRICIAN
Practice Exam #8

The following questions are based on the 2008 edition of the National Electrical Code® and are typical of questions encountered on most Journeyman Electricians' Exams. Select the best answer from the choices given and review your answers with the answer key included in this book.

ALLOTTED TIME: 75 minutes

Notes

Journeyman Electrician Practice Exam #8

1. In general, motor control circuits shall be arranged so that they will be disconnected from all sources of supply when _____.

   a. the disconnecting means is in the open position.
   b. the disconnecting means is not within sight of the controller
   c. the disconnecting means is integral with the controller
   d. installed in a motor control center

2. Which of the following is permitted for closing an unused circuit breaker opening in panelboards?

   a. Installation of a closure plate or other device that fits securely and closes the opening.
   b. Installation of a spare or unused circuit breaker that fills the opening and is listed for the panelboard.
   c. Installation of a closure plate listed for the purpose by the manufacturer.
   d. All of these are permitted.

3. Before demand factors are taken into consideration for commercial buildings, general purpose receptacle loads are to be computed at not less than _____ VA per single or duplex receptacle.

   a. 100
   b. 120
   c. 150
   d. 180

4. Which of the following listed conductor insulations have a greater temperature rating when used in a dry location compared to when used in a wet location?

   a. THW
   b. RHW
   c. THWN
   d. THHW

5. Which of the following situations regarding grounding, would NOT be in compliance with the standards set forth by the NEC®?

   I. Steel reinforcing bars to be poured in a concrete foundation and not connected as part of the grounding system.

   II. Concrete foundations with steel reinforcing bars that were not available for use as a grounding electrode because the concrete was poured before the electrical contractor arrived on the job.

   a. I only
   b. II only
   c. both I and II
   d. neither I nor II

6. In a health care facility, low-voltage electrical equipment that is likely to become energized and is frequently in contact with the bodies of patients shall operate on a voltage of _____ volts or less if the equipment is not approved as intrinsically safe, double-insulated, or moisture-resistant.

   a. 10
   b. 24
   c. 100
   d. 120

7. When ambient temperature is not a factor, a size 8 AWG single copper conductor with type FEPB insulation, installed in free air, will have a MAXIMUM allowable ampacity of _____.

   a. 80 amperes
   b. 55 amperes
   c. 45 amperes
   d. 83 amperes

Notes

Notes

8. For the purpose of determining conductor fill in a device box, a switch is counted as equal to _____ conductor(s), based on the largest conductor connected to the switch.

   a. no
   b. one
   c. two
   d. three

9. When determining the required ampacity of the branch-circuit conductors to supply an AC three-phase, continuous-duty motor, current values of the motor given on _____ shall be used.

   a. the motors nameplate
   b. Table 430.248
   c. Table 430.52
   d. Table 430.250

10. When calculating the total load for a commercial building, what is the MINIMUM computed load, in volt-amps, required by the NEC® for a branch-circuit supplying an exterior sign?

    a. 1,200 volt-amps
    b. 1,800 volt-amps
    c. 1,500 volt-amps
    d. 2,000 volt-amps

11. Determine the MAXIMUM number of size 1 AWG XHHW compact conductors permitted in a trade size 3 inch, Schedule 40 rigid PVC conduit that is 100 feet long.

    a. 21
    b. 19
    c. 14
    d. 18

12. The MINIMUM headroom of working spaces about service equipment and motor control centers shall be at LEAST _____.

    a. 6 feet
    b. 6 feet, 6 inches
    c. 6 feet, 7 inches
    d. 8 feet

13. Class _____ hazardous locations are those in which flammable gases, flammable liquid-produced vapors, or combustible liquid-produced vapors are or may be present in the air in quantities sufficient to produce explosive or ignitible mixtures.

    a. I
    b. II
    c. III
    d. IV

14. In indoor areas where walls are frequently washed, such as laundries or car washes, metal conduit and metal panelboards shall be mounted with a _____ space between the wall and conduit or panelboard when the equipment is installed exposed.

    a. ¾ inch
    b. ½ inch
    c. ⅛ inch
    d. ¼ inch

15. A branch-circuit supplied by size 12 AWG conductors is a _____ rated branch-circuit when protected by a 15 ampere rated circuit breaker.

    a. 20-ampere
    b. 25-ampere
    c. 15-ampere
    d. 10-ampere

Notes

Notes

16. For other than single-family dwellings, spas and hot tubs are required to have a clearly labeled emergency shutoff or control switch provided. This switch shall be installed at a point readily accessible to the users and not less than _____ feet and within sight of the spa or hot tub.

    a. 12
    b. 10
    c. 15
    d. 5

17. In a data-processing room, the disconnecting means for the data-processing equipment shall be _____.

    a. within sight of the equipment
    b. near an exit door
    c. at the main disconnect
    d. within 30 feet of the equipment

18. Type MI cable shall be supported at intervals not exceeding _____ feet.

    a. 2
    b. 4
    c. 6
    d. 10

19. The internal depth of an outlet box used only to splice conductors to a luminaire shall NOT be less than _____.

    a. ½ inch
    b. 15⁄16 inch
    c. ¾ inch
    d. 1 inch

20. Direct-buried conductors emerging from grade shall be protected by raceways or covers up to at LEAST _____ above finished grade.

    a. 8 feet
    b. 10 feet
    c. 12 feet
    d. 15 feet

21. What is the MINIMUM size equipment grounding conductor required by the NEC® for a branch-circuit protected by a 50-ampere rated circuit breaker?

    a. 12 AWG
    b. 10 AWG
    c. 8 AWG
    d. 6 AWG

22. Disregarding exceptions, all three-phase, 277/480 volt, wye-connected electrical services require ground-fault protection for each service disconnecting means when rated at LEAST _____ or more.

    a. 400 amperes
    b. 600 amperes
    c. 1,000 amperes
    d. 1,500 amperes

Notes

Notes

23. Determine the conductor ampacity given the following related information:

    - conductors are size 500 kcmil copper
    - conductor insulation is THWN
    - eight current-carrying conductors are in the raceway
    - ambient temperature is 125°F
    - conduit length is 75 feet
    - installation is a wet location

    a. 178.2 amperes
    b. 199.5 amperes
    c. 294.6 amperes
    d. 380.0 amperes

24. Flexible metal conduit (FMC) is permitted for use in wet locations _____.

    a. when the conductors in the FMC have a temperature rating of 90°C
    b. when the conductors contained within the FMC are approved for use in a wet location
    c. if the length of the FMC is not more than 6 feet
    d. never

25. Given: a trade size ½-inch rigid metal conduit (RMC) is to be installed in a Class I location and a trade size ½ conduit seal is required. The MINIMUM thickness of the sealing compound shall NOT be less than _____.

    a. ⅜ inch
    b. ½ inch
    c. ⅝ inch
    d. ¾ inch

JOURNEYMAN ELECTRICIAN
Practice Exam #9

The following are based on the 2008 edition of the National Electrical Code® and are typical of questions encountered on most Journeyman Electricians' Exams. Select the best answer from the choices given and review your answers with the answer key included in this book.

ALLOTTED TIME: 75 minutes

Notes

Journeyman Electrician Practice Exam #9

1. Where a water pipe passes over an indoor motor control center, leak protection apparatus shall have a MINIMUM clearance of _____.

   a. 6 feet to the protection apparatus
   b. 6 feet to the water pipe
   c. 6½ feet to the protection apparatus
   d. 6½ feet to the water pipe

2. Branch circuit grounded conductors that are not connected to an overcurrent device shall be permitted to be sized at LEAST _____ of the continuous and noncontinuous load to be served.

   a. 100 percent
   b. 80 percent
   c. 75 percent
   d. 125 percent

3. The MAXIMUM standard rating of the initial non-time-delay fuse for the branch-circuit, short-circuit, and ground-fault protection of a 1 HP, three-phase, 208-volt, squirrel-cage AC motor shall be _____.

   a. 6 amperes
   b. 10 amperes
   c. 15 amperes
   d. 20 amperes

4. For Class III equipment that is not subject to overloading the MAXIMUM surface temperature under operating conditions shall NOT exceed _____.

   a. 120°C
   b. 165°C
   c. 248°F
   d. 210°C

5. When a metal wireway contains splices and taps, the conductors, splices and taps shall not fill the wireway to more than _____ of its area at that point.

   a. 20 percent
   b. 30 percent
   c. 40 percent
   d. 75 percent

6. In an underground installed Schedule 40 PVC conduit system that consists of 50 feet in length between pulling points, what is the MAXIMUM number of bends this conduit run may have?

   a. four 90° bends
   b. six 90° bends
   c. four 120° bends
   d. two 90° bends

7. Grounding electrodes made of pipe or conduit shall not be smaller than _____ electrical trade size.

   a. ½ inch
   b. ¾ inch
   c. 1 inch
   d. 1½ inch

8. Liquid-tight flexible metallic conduit (LFMC) shall NOT be used _____.

   a. in lengths in excess of 6 feet
   b. in concealed work
   c. in hazardous locations
   d. where subject to physical damage

Notes

Notes

9. Determine the allowable ampacity of a size 3 AWG THHN copper conductor given the following:

   - Two current-carrying conductors are in the conduit.
   - The ambient temperature is 34°C.
   - The terminations are rated 60°C.

   a. 105.6 amperes
   b. 85.0 amperes
   c. 81.6 amperes
   d. 110 amperes

10. When an overcurrent protective device for a surge-protective device is located at the service location it _____.

    a. is not to be counted as a service disconnect
    b. shall be considered as a service disconnecting means
    c. shall be more than 10 feet from the main disconnecting means
    d. shall be prohibited

11. Standard trade size 2-inch rigid polyvinyl chloride conduit (PVC), shall be supported at least every _____.

    a. 3 feet
    b. 5 feet
    c. 6 feet
    d. 8 feet

12. When sizing time-delay Class CC fuses for motor branch-circuit, short-circuit, and ground-fault protection, they are to be sized at the same value as _____.

    a. inverse-time circuit breakers
    b. non-time-delay fuses
    c. instantaneous trip circuit breakers
    d. adjustable trip circuit breakers

13. Where sizes 1/0 AWG through 4/0 AWG single conductor cables are installed in ladder type cable tray, the MAXIMUM allowable rung spacing for the ladder cable tray shall be _____.

    a. 6 inches
    b. 9 inches
    c. 12 inches
    d. 15 inches

14. The NEC® requires the MAXIMUM length permitted for a flexible cord supplying a 208-volt, single-phase, room air-conditioner to be _____.

    a. 4 feet
    b. 6 feet
    c. 8 feet
    d. 10 feet

15. In general, which of the following receptacle outlets in a commercial kitchen are required by the NEC® to be GFCI protected?

    a. All 125-volt, single-phase, 15- and 20-ampere rated receptacles.
    b. All 125-volt, 15- or 20-ampere rated receptacles in wet locations only.
    c. All receptacle outlets.
    d. All 125-volt countertop receptacle outlets only.

16. All wiring installed in or under an aircraft hanger floor shall comply with the requirements for _____ locations.

    a. Class I, Division 1
    b. Class I, Division 2
    c. Class II, Division 1
    d. Class II, Division 2

Notes

Notes

17. In general, branch-circuit conductors supplying more than one motor shall have an ampacity of at least _____ percent of the FLC of the largest motor, and 100 percent of the FLC of the other motor(s) in the group.

    a. 25
    b. 80
    c. 100
    d. 125

18. What is the MINIMUM size equipment grounding conductor required for a 5 HP, three-phase, 208-volt motor having 20-ampere rated overload protection and short-circuit overcurrent protection rated at 30 amperes?

    a. 10 AWG
    b. 14 AWG
    c. 12 AWG
    d. 8 AWG

19. If a circuit breaker serves as the controller for a motor, and the motor is not in sight of the breaker, the NEC® requires which of the following?

    a. The motor to be less than 2 HP.
    b. The breaker be able to be locked in the open position.
    c. The motor to be Code letter "E".
    d. The breaker to be rated 25,000 AIC.

20. In Class II, Division 1, hazardous locations, an approved method of connection of conduit to boxes is _____.

    a. compression fittings
    b. threaded bosses
    c. welding
    d. all of these

21. Given: an AC service is supplied with four parallel sets of size 500 kcmil aluminum conductors. What is the MINIMUM size copper grounding electrode required, when connected to the concrete-encased building steel used as the grounding electrode system?

    a. 3/0 AWG
    b. 4/0 AWG
    c. 250 kcmil
    d. 2/0 AWG

22. Given: an AC transformer arc welder has a 50-ampere rated primary current and a 60 percent duty-cycle. Determine the MINIMUM size copper 60°C rated conductors the NEC® requires to supply this welder.

    a. 6 AWG
    b. 8 AWG
    c. 10 AWG
    d. 4 AWG

23. When installing wiring for sensitive electronic equipment, the MAXIMUM voltage to ground is required to be _____ volts.

    a. 30
    b. 60
    c. 120
    d. 300

24. In a health care facility, a patient bed location in a critical care area is required to have which of the following?

    a. four single receptacles or two duplex receptacles
    b. six single receptacles or three duplex receptacles
    c. two duplex receptacles or four single receptacles
    d. two single receptacles or one duplex receptacle

Notes

Notes

25. Which of the following circuits is prohibited for the grounded conductor to be dependent on receptacle devices for continuity?

    a. all circuits

    b. multi-outlet circuits

    c. GFCI protected circuits

    d. multi-wire circuits

JOURNEYMAN ELECTRICIAN
Practice Exam #10

The following questions are based on the 2008 edition of the National Electrical Code® and are typical of questions encountered on most Journeyman Electricians' Exams. Select the best answer from choices given and review your answers with the answer key included in this book.

ALLOTTED TIME: 75 minutes

Notes

Journeyman Electrician Practice Exam #10

1. Where conduits enter a floor-standing switchboard or panelboard at the bottom, the conduits including their end fittings, shall not rise more than _____ above the bottom of the enclosure.

   a. 6 inches
   b. 4 inches
   c. 2 inches
   d. 3 inches

2. When installed on the service of a building or structure, a type 2 SPD (TVSS) shall be connected _____.

   a. on the line side of the service disconnect overcurrent device
   b. on the load side of the service disconnect overcurrent device
   c. on the meter housing
   d. at the final overcurrent device

3. A detached garage supplied from a dwelling with a three-wire feeder of size 4/0 AWG aluminum USE cable with a 200-ampere rated circuit breaker for overcurrent protection requires a MINIMUM size copper grounding electrode to a concrete encased electrode of _____ AWG.

   a. 8
   b. 6
   c. 4
   d. 2

4. When a raceway is run from the interior to the exterior of a building, the raceway shall _____.

   a. be rigid metal conduit (RMC)
   b. be rigid polyvinyl chloride conduit (PVC)
   c. be sealed with an approved material
   d. have an explosion-proof seal

5. A disconnecting means shall be provided on the supply side of all fuses in circuits having a voltage of at LEAST _____ volts to ground.

   a. 120
   b. 250
   c. 150
   d. 277

6. Given: an 80-ampere, 240-volt, single-phase, load is located 200 feet from the supply panelboard and is supplied with size 3 AWG copper conductors with THHN/THWN insulation. What is the approximate voltage-drop on this circuit? (K = 12.9)

   a. 6.0 volts
   b. 4.0 volts
   c. 9.2 volts
   d. 7.84 volts

7. Enclosures in a Class I, Division 1, location containing components that have arcing devices must have an approved seal located within _____ of each conduit run entering or leaving such enclosures.

   a. 12 inches
   b. 18 inches
   c. 24 inches
   d. 30 inches

8. Surface-mounted incandescent luminaires installed on the wall above the door or on the ceiling of a clothes closet shall have a clearance of at LEAST ____ between the luminaire and the nearest point of storage.

   a. 6 inches
   b. 8 inches
   c. 18 inches
   d. 12 inches

Notes

Notes

9. The NEC® considers the area around the fuel dispensing pumps at a service station to be a hazardous location. This area extends a height of 18 inches above grade, and up to a distance from the enclosure of the dispensing pump of _____.

    a. 5 feet
    b. 10 feet
    c. 16 feet
    d. 20 feet

10. According to the National Electrical Code®, what are the two mandatory branches of the hospital emergency system?

    a. The emergency branch and the standby branch.
    b. The life safety branch and the critical branch.
    c. The normal branch and the alternate branch.
    d. The essential branch and the non-essential branch.

11. Class 3 single-conductors shall NOT be smaller than _____.

    a. 18 AWG
    b. 16 AWG
    c. 14 AWG
    d. 20 AWG

12. Given: an office building has a 208Y/120 volt, three-phase service with a balanced net computed load of 90 kVA. What is the current each ungrounded conductor carries at full load?

    a. 188 amperes
    b. 250 amperes
    c. 433 amperes
    d. 750 amperes

13. When grounding the structural reinforcing steel of a swimming pool, what is the smallest size grounding conductor permitted for this installation?

   a. 12 AWG
   b. 10 AWG
   c. 8 AWG
   d. 6 AWG

14. A surge-protective device (SPD) is a protective device for _____.

   a. limiting transient voltage by diverting or limiting surge current
   b. limiting transient voltage by absorbing and storing excessive surge current
   c. switching transient voltage back to the line side of the service transformer
   d. none of these

15. Storage batteries used as a source of power for emergency systems shall be of a suitable rating and capacity to supply and maintain the total load for at LEAST _____.

   a. 1/2 hour
   b. 1 hour
   c. 1 1/2 hours
   d. 2 hours

16. Given: a size 14 AWG branch-circuit conductor protected by a 15-ampere rated circuit breaker is to serve three 20-ampere rated duplex receptacles. This branch-circuit:

   a. would be in compliance with the NEC® if the wire was size 12 AWG.
   b. would be in compliance with the NEC® if the circuit breaker was rated at 20 amperes.
   c. is not in compliance with the NEC® because the receptacle outlets have a 20-ampere rating.
   d. is in compliance with the NEC®.

Notes

Notes

17. Where fuses are used for motor overload protection, if the supply system is a three-wire, three-phase, AC with one conductor grounded, the NEC® requires at LEAST _____ fuse(s).

    a. zero
    b. one
    c. two
    d. three

18. Given: an existing metal junction box has a volume of 27 cubic inches and contains a total of six size 12 AWG wires. Additional wires of size 10 AWG need to be added in the box. No grounding conductors, devices or fittings are contained in the box. What is the MAXIMUM number of size 10 AWG conductors that may be added to this box?

    a. two
    b. five
    c. six
    d. eight

19. For enclosing a mobile home supply cord, what is the MAXIMUM trade size conduit permitted between the branch-circuit panelboard of a mobile home and the underside of a mobile home floor?

    a. 1 inch
    b. 1¼ inch
    c. 1½ inch
    d. 2 inch

20. Which of the following wiring methods is/are approved for use for fixed wiring in an area above Class I locations in a commercial garage?

    a. type MI cable
    b. type TC cable
    c. type MC cable
    d. all of the above

21. Class II locations are those that are hazardous (classified) because of the presence of _____.

    a. flammable gas
    b. ignitible fibers
    c. ignitible vapors
    d. combustible dust

22. Branch circuits supplying a fixed, storage-type, electric water heater with a capacity of 120 gallons or less, shall NOT be less than _____ of the full-load current of the water heater.

    a. 80 percent
    b. 100 percent
    c. 125 percent
    d. 150 percent

23. Household-type appliances with surface heating elements having a MAXIMUM demand of more than _____ amperes, shall have its power supply subdivided into two or more circuits.

    a. 60
    b. 50
    c. 40
    d. 30

24. When driving a ground rod and solid rock is encountered, the ground rod is permitted to be buried in a trench that is at LEAST _____ deep.

    a. 3 feet
    b. 4 feet
    c. 24 inches
    d. 30 inches

Notes

Notes

25. Ten copper THW conductors are to be installed in a 20-foot length of intermediate metal conduit (IMC); five size 1 AWG and five size 3 AWG. What is the MINIMUM allowable trade size IMC required to contain these conductors?

    a. 1½ inches

    b. 2 inches

    c. 2½ inches

    d. 4 inches

JOURNEYMAN ELECTRICIAN
Practice Exam #11

The following questions are based on the 2008 edition of the National Electrical Code® and are typical of questions encountered on most Journeyman Electricians' Exams. Select the best answer from the choices given and review your answers with the answer key included in this book.

ALLOTTED TIME: 75 minutes

Notes

Journeyman Electrician
Practice Exam #11

1. Unless listed and identified otherwise, terminations and conductors of sizes 14 AWG through 1 AWG shall be rated at _____.

   a. 75°C
   b. 75°F
   c. 90°C
   d. 60°C

2. The MINIMUM size copper equipment bonding jumper on the supply side of a service consisting of parallel copper phase conductors with a total circular mil area of 2000 kcmil shall be ____.

   a. 2/0 AWG
   b. 3/0 AWG
   c. 4/0 AWG
   d. 250 kcmil

3. Which of the following grounding electrodes is NOT permitted to be used as an effective ground-fault current path?

   a. the metal frame of a building or structure
   b. a metal underground water pipe in direct contact with the earth for 10 feet or more
   c. an electrode encased by at least 2 inches of concrete
   d. the earth

4. Primary overcurrent devices, impedance limiting means, or other inherent protective means shall be permitted to protect a motor control circuit transformer when the transformer is rated no more than _____ VA.

   a. 50
   b. 60
   c. 75
   d. 100

5. When installing electric space heating cables in ceilings, which of the following statements is/are true?

   a. The heating cables are permitted to be installed over walls.
   b. The non-heating leads of the cables shall be at least 6 inches in length.
   c. The heating cables shall be kept free from contact with other electrically conductive surfaces.
   d. All of these statements are true.

6. Given: a unbroken length of size 12 AWG conductor is looped inside a junction box so that there is more than 12 inches of free conductor. The conductor is not spliced and does not terminate on a device. What volume, in cubic inches, must be counted for the looped conductor?

   a. 2.25 cubic inches
   b. 3.25 cubic inches
   c. 4.50 cubic inches
   d. 4.00 cubic inches

7. Nonmetallic sheathed cable is permitted to be used _____.

   a. in buildings of type III, IV, and V construction
   b. as open runs in suspended ceilings of commercial buildings
   c. in buildings of type I and II construction
   d. in conduit installed underground

8. Given: a boat repair shop has several 120-volt, 20-ampere general-purpose receptacle outlets installed at a height of 16 inches in a work area where gasoline may be handled and mechanical ventilation is not provided. The outlets are considered to be installed in a/an _____ location.

   a. Class I, Division 1
   b. Class I, Division 2
   c. Class II, Division 1
   d. Unclassified

Notes

Notes

9. A luminaire rated at 7 amperes requires a MINIMUM size _____ AWG fixture wire.

   a. 12
   b. 14
   c. 16
   d. 18

10. All receptacle outlets located in areas of pediatric wards of hospitals are required to _____.

    I. be listed tamper-resistant
    II. employ a listed tamper-resistant cover

    a. I only
    b. II only
    c. either I or II
    d. neither I nor II

11. An installation of 40 feet of electrical metallic tubing (EMT) will contain the following conductors:

    - Three size 2 AWG with THWN insulation
    - Four size 6 AWG with THWN insulation
    - Three size 10 AWG with THWN insulation

    This installation requires a MINIMUM trade size _____ EMT.

    a. 1¼ inch
    b. 1½ inch
    c. 2 inch
    d. 2½ inch

12. Where UF cable is used as a substitute for Type NM cable, the conductor insulation is required to be rated _____.

    a. 60°C
    b. 75°C
    c. 75°F
    d. 90°C

13. When armored cable (AC) is installed in thermal insulation, the ampacity of the cable shall be that of _____ conductors.

    a. 60°C
    b. 75°C
    c. 194°F
    d. 90°C

14. Receptacle outlets installed on a branch-circuit rated at 30 amperes are permitted to have an ampere rating of _____.

    a. 15, 20, or 30 amperes
    b. 20 or 30 amperes
    c. 30 or 40 amperes
    d. 30 amperes only

15. When conductors or cables are installed in conduits exposed to direct sunlight on or above a rooftop and when they are within at LEAST _____ of the rooftop a temperature adder must be applied to the applicable correction factors of Table 310.16 to determine the allowable ampacity of the conductors.

    a. 48 inches
    b. 36 inches
    c. 24 inches
    d. 12 inches

16. Disregarding exceptions, where a feeder supplies continuous loads or any combination of continuous and non-continuous loads, the rating of the overcurrent device shall NOT be less than the non-continuous loads plus _____ of the continuous loads.

    a. 80 percent
    b. 100 percent
    c. 125 percent
    d. 150 percent

Notes

Notes

17. Surge-protective devices (SPDs) and transient voltage surge suppressors (TVSSs) may be installed on circuits NOT exceeding _____ volts.

    a. 150
    b. 300
    c. 600
    d. 1,000

18. When buried raceways pass under a driveway, the MINIMUM cover requirements _____.

    a. decrease if installed in rigid metal conduit (RMC)
    b. do not change in regard to wiring methods used
    c. shall be increased for direct buried cables
    d. can be increased, decreased, or remain the same, depending on the wiring method used

19. Exposed nonmetallic surface extensions shall be permitted to run in any direction from an existing outlet, but not within _____ from the floor.

    a. 1 foot
    b. 1½ feet
    c. 2 feet
    d. 2 inches

20. Which of the following listed luminaires is NOT permitted for installations in clothes closets?

    a. recessed incandescent luminaire
    b. surface-mounted fluorescent luminaire
    c. pendant luminaire
    d. surface-mounted incandescent luminaire

21. Given: an office building contains 200 general purpose, 120-volt duplex receptacle outlets rated at 15 amperes each. What is the total connected load, in VA, for the receptacles?

    a. 36,000 VA
    b. 18,000 VA
    c. 54,000 VA
    d. 72,000 VA

22. Which of the following circuit breakers are NOT permitted to be installed in a 120/240-volt, single-phase, power panel?

    a. double-pole circuit breakers
    b. bolt-in style circuit breaker
    c. indicating-type circuit breakers
    d. delta-style circuit breakers

23. Regardless of the voltage, the MINIMUM clearance above a diving platform of a swimming pool and service drop conductors is _____.

    a. 14½ feet
    b. 22½ feet
    c. 19 feet
    d. 21 feet

24. Given: a small retail store will have an area of 1,600 square feet. Disregarding exceptions, how many 15-ampere, 120-volt general lighting branch-circuits are required for this installation?

    a. one
    b. two
    c. three
    d. four

Notes

Notes

25. An aluminum grounding electrode shall be _____.

    a. installed at least 18 inches above the earth

    b. installed at least 24 inches above the earth

    c. in direct earth burial

    d. prohibited

MASTER ELECTRICIAN
Practice Exam #12

The following questions are based on the 2008 edition of the National Electrical Code® and are typical of questions encountered on most Master Electricians' Exams. Select the best answer from the choices given and review your answers with the answer key included in this book.

ALLOTTED TIME: 75 minutes

Notes

Master Electrician
Practice Exam #12

1. The MINIMUM size copper grounding electrode conductor tap for a separately derived alternating current electrical system with three paralleled sets of size 300 kcmil copper secondary conductors shall be _____.

   a. 1/0 AWG
   b. 1 AWG
   c. 2/0 AWG
   d. 3/0 AWG

2. Cables or raceways installed using directional boring equipment shall _____.

   a. be approved for the purpose
   b. have a warning ribbon installed above the cable or raceway
   c. not be permitted
   d. be identified by color coding

3. Given: power for equipment that is directly associated with the radio frequency distribution system is carried by the coaxial cable. The power source is a power limiting transformer. The MAXIMUM voltage this coaxial cable may carry is _____ volts.

   a. 50
   b. 60
   c. 120
   d. 150

4. Given: a 15-ampere rated general-use AC snap switch is to serve as a disconnecting means for an AC electric motor. The NEC® requires the MAXIMUM full-load current rating of the motor to be no more than _____.

   a. 7.5 amperes
   b. 10 amperes
   c. 12 amperes
   d. 15 amperes

5. In general, the MINIMUM voltage allowed at the line terminals of a 480-volt, three-phase, fire pump controller under motor starting conditions shall be _____.

    a. 466 volts
    b. 456 volts
    c. 432 volts
    d. 408 volts

6. Which of the following statements, if any, is/are true regarding the illumination for service equipment installed in electrical equipment rooms of commercial or industrial occupancies?

    I. The illumination shall not be controlled by means of three-way switches.

    II. The illumination shall not be controlled by automatic means only.

    a. I only
    b. II only
    c. both I and II
    d. neither I nor II

7. Given: you have tenant spaces in a retail shopping mall; each occupant shall have access to the main disconnecting means, EXCEPT _____.

    a. where the service and maintenance are provided by the building management
    b. where there are more than six disconnecting means provided
    c. where the primary feeder transformer does not exceed 600 volts
    d. where the secondary of the service transformer does not exceed 240 volts to ground

Notes

Notes

8. Determine the conductor allowable ampacity given the following related information:

   - ambient temperature of 44°C
   - 250 kcmil THWN copper conductors
   - four current-carrying conductors in the raceway
   - length of raceway is 30 feet

   a. 160 amperes
   b. 167 amperes
   c. 200 amperes
   d. 209 amperes

9. Determine the MAXIMUM overcurrent protection permitted for size 14 AWG THWN copper motor control circuit conductors tapped from the load side of a motor overcurrent protection device.

   Given: the conductors require short-circuit protection and do not extend beyond the motor control equipment enclosure.

   a. 20 amperes
   b. 25 amperes
   c. 30 amperes
   d. 100 amperes

10. Which of the following statement(s) is/are TRUE regarding grounding and bonding of metal gas lines?

    a. A buried natural gas line may be used as an electrical system only grounding conductor.
    b. A buried natural gas line may be used as a grounding electrode when supplemented by another electrode.
    c. The furnace gas line must be bonded to an electrical grounding system.
    d. All of these are true.

11. Given: a grounding ring consist of bare copper wire encircling a building, buried 36 inches, in direct contact with the earth. The NEC® requires the MINIMUM size of the wire to the grounding electrode in this ground ring to be _____.

    a. 2 AWG
    b. 3 AWG
    c. 4 AWG
    d. 6 AWG

12. All 15- and 20-ampere, 125-volt, single-phase, nonlocking receptacles installed outdoors in wet locations of a commercial establishment shall _____.

    I. be a listed weather-resistant type
    II. have a ground-fault circuit interrupter protection for personal

    a. I only
    b. II only
    c. both I and II
    d. neither I nor II

13. Which one of the following conductor insulations is identified as oil-resistant?

    a. TW
    b. TFE
    c. THWN
    d. MTW

14. Class A ground-fault circuit interrupters are designed to trip when the current to ground is at LEAST _____ or higher.

    a. 6 mA
    b. 4 mA
    c. 8 mA
    d. 10 mA

Notes

Notes

15. Determine the MINIMUM number of 15-ampere, 120-volt, general lighting branch-circuits required for a 12,000 square feet multi-family dwelling when each unit has cooking facilities provided.

    a. 15
    b. 20
    c. 24
    d. 30

16. At least one receptacle outlet shall be installed within 18 inches of the top of a show window of a retail department store for each _____ linear feet of show window lighting area.

    a. 6
    b. 8
    c. 10
    d. 12

17. When a conduit containing service-entrance conductors runs beneath a building, what is the MINIMUM depth of concrete required to cover a conduit for it to be considered "outside" the building?

    a. 2 inches
    b. 6 inches
    c. 12 inches
    d. 18 inches

18. Continuous-duty motors with a marked service factor not less than 1.15 shall have the MINIMUM running overload protection sized at _____ percent of the nameplate rating of the motor.

    a. 115
    b. 125
    c. 130
    d. 140

19. A 90-ampere, 240-volt, single-phase load is located 225 feet from a panelboard and is supplied with size 3 AWG THWN copper conductors. What is the approximate voltage-drop on this circuit? (K = 12.9)

    a. 6 volts
    b. 4 volts
    c. 8 volts
    d. 10 volts

20. A single-family residence is served with insulated 120/240 volt service-drop conductors, supported with a grounded messenger wire from the utility company. This service-drop crosses the swimming pool; the pool has a diving board that is 5 feet above the water level. The MINIMUM height above water level of the service drop conductors as permitted by the NEC® is _____.

    a. 25 feet
    b. 14½ feet
    c. 22½ feet
    d. 19½ feet

21. At an outdoor motor fuel dispensing facility, the area up to 18 inches above grade and within 20 feet from the edge of a LPG dispenser enclosure shall be classified as _____.

    a. Class I, Division 1
    b. Class I, Division 2
    c. Class II, Division 1
    d. Class II, Division 2

22. In a commercial garage work area, which of the following receptacles, if any, are required to have GFCI protection?

    I. 15-ampere, 120-volt general purpose receptacles for hand tools and portable lights.
    II. 20-ampere, 120-volt receptacles serving electrical diagnostic equipment only.

    a. I only
    b. II only
    c. both I and II
    d. neither I nor II

Notes

Notes

23. Given: a feeder is to supply two continuous-duty, induction type, Design B, 208-volt, three-phase AC motors, one 10 HP and one 7½ HP; the feeder shall have a MINIMUM ampacity of _____.

    a. 55.00 amperes
    b. 68.75 amperes
    c. 62.70 amperes
    d. 38.50 amperes

24. The MINIMUM spacing required between the bottom of a 600-volt rated switchboard and noninsulated busbars mounted in the switchboard cabinet is _____.

    a. 6 inches
    b. 8 inches
    c. 10 inches
    d. 12 inches

25. Given: A rigid metal conduit (RMC) contains only the following three circuits on the load side of the service overcurrent protection devices:

    - two 150-ampere, three phase-circuits
    - one 300-ampere single phase-circuit

    The load side equipment bonding jumper for this conduit must be a MINIMUM size _____ copper.

    a. 1 AWG
    b. 2 AWG
    c. 4 AWG
    d. 6 AWG

MASTER ELECTRICIAN
Practice Exam #13

The following questions are based on the 2008 edition of the National Electrical Code® and are typical of questions encountered on most Master Electricians' Exams. Select the best answer from the choices given and review your answers with the answer key included in this book.

ALLOTTED TIME: 75 minutes

Notes

Master Electrician
Practice Exam #13

1. Given: a hotel has a stationary electric sign mounted inside a large decorative fountain. The sign shall be at LEAST _____ feet inside the fountain, when measuring from the outside edges of the fountain.

   a. 2
   b. 3
   c. 4
   d. 5

2. Which one of the following statements regarding handhole enclosures for use in underground sytems is correct?

   a. The enclosure must be sealed to prevent moisture from entering the enclosure.
   b. The enclosure is mounted on a pedestal.
   c. The enclosure must have a hinged cover.
   d. The enclosure must be large enough to permit access by reaching into the enclosure.

3. Branch-circuit conductors supplying outlets for professional type arc and Xenon motion picture projectors shall be at LEAST size _____.

   a. 12 AWG
   b. 10 AWG
   c. 8 AWG
   d. 6 AWG

Notes

4. Where rigid nonmetallic conduit (RNC) is installed underground to supply a gasoline dispensing unit, threaded rigid metal conduit (RMC) or threaded intermediate metal conduit (IMC) shall be used for the last _____ feet of the underground run to where the conduit emerges.

   a. 2
   b. 4
   c. 5
   d. 6

5. Given: a feeder tap conductor is to terminate in a 70-ampere rated circuit breaker. The ampacity of the feeder tap conductor is required to be not less than ______.

   a. 65 amperes
   b. 70 amperes
   c. 85 amperes
   d. 140 amperes

6. When determining the number of conductors permitted to be installed in an outlet box, a reduction of _____ conductors shall be made for, one device, two clamps, and three grounding conductors.

   a. three
   b. four
   c. five
   d. six

7. The demand factor of any electrical system is the ratio of the maximum demand for a system to _____.

   a. the total connected load of the system
   b. 125 percent of the total connected load of the system
   c. 125 percent of the total connected continuous load of the system
   d. 80 percent of the total connected noncontinuous load plus 125 percent of the total connected continuous load of the system

Notes

8. All 15- or 20-ampere, single-phase, 125-volt receptacle outlets located within at LEAST _____ of the edge of a decorative fountain shall be provided with GFCI protection.

   a. 10 feet
   b. 15 feet
   c. 20 feet
   d. 25 feet

9. The emergency controls for attended self-service gasoline stations and convenience stores must be located NOT more than _____ from the gasoline dispensers.

   a. 20 feet
   b. 50 feet
   c. 75 feet
   d. 100 feet

10. Hospital x-ray equipment supplied by a branch-circuit rated at NOT more than _____ amperes may be served by a hard-service cord with a suitable attachment plug.

   a. 15
   b. 20
   c. 30
   d. 60

11. What are the MINIMUM size THWN copper conductors required to serve a continuous-duty, 25 HP, 208-volt, three-phase motor?

    Given: the motor is on the end of a 25-foot conduit run that contains only three conductors at an expected ambient temperature of 50°C.

   a. 6 AWG
   b. 3 AWG
   c. 2 AWG
   d. 1 AWG

12. Outlets supplying permanently installed swimming pool pump motors from single-phase 15- or 20-ampere, 120- or 240-volt branch circuits shall be provided with GFCI protection _____.

    a. when installed in buildings
    b. when cord-and-plug connected
    c. when direct connected
    d. always

13. Which one of the following listed shall be installed on the critical branch of the emergency system in a health care facility?

    a. exit signs
    b. nurse call systems
    c. communication systems
    d. fire alarms

14. A disconnecting means installed at the distribution point where two or more agricultural buildings are supplied shall NOT be required to _____.

    a. be grounded
    b. contain overcurrent protection
    c. contain a grounded conductor
    d. be rated for the calculated load

15. When calculating the service-entrance conductors for a farm service, the second largest load of the total load, shall be computed at not less than _____ percent.

    a. 90
    b. 80
    c. 75
    d. 65

Notes

Notes

16. When a building contains multiple elevators, a lighting branch-circuit supplying the cars may supply no more than _____ car(s).

    a. one
    b. two
    c. three
    d. four

17. What percent of electrical supplied spaces in a recreational vehicle park must be provided with 30-ampere, 125-volt receptacle outlets?

    a. 60
    b. 70
    c. 90
    d. 100

18. The collector ring used for grounding of an irrigation machine shall have a current rating of NOT less than _____ of the full-load current of the largest device served plus the sum of the other devices served.

    a. 75 percent
    b. 80 percent
    c. 100 percent
    d. 125 percent

19. Given: a three-phase, 25 KVA rated transformer with a 480-volt primary and a 208Y/120 volt secondary is to be installed. Where primary and secondary protection is required to be provided, what is the MAXIMUM standard ampere rating of secondary overcurrent protection permitted by the NEC®?

    a. 80 amperes
    b. 90 amperes
    c. 100 amperes
    d. 110 amperes

20. Where ungrounded conductors are run in parallel in multiple raceways, the equipment grounding conductor, where used, shall be _____.

    a. omitted
    b. run in parallel
    c. installed in one raceway only
    d. bare

21. In agricultural buildings where livestock is housed, any portion of a direct-buried equipment grounding conductor run to the building or structure shall be _____.

    a. insulated or covered copper
    b. bare or insulated copper
    c. bare or insulated aluminum
    d. insulated or covered copper or aluminum

22. Determine the MINIMUM size THWN copper feeder conductors required by the NEC® to supply the following 480-volt, continuous-duty, three-phase, induction-type, Design B, AC motors.

    - one 40 HP
    - one 50 HP
    - one 60 HP

    a. 2/0 AWG
    b. 3/0 AWG
    c. 4/0 AWG
    d. 250 kcmil

Notes

Notes

23. Two or more motors may be installed without individual overcurrent protection devices if rated less than 1 HP each; and the full load current rating of each motor does NOT exceed _____ amperes, if they are on a 120-volt, single-phase, 20-ampere rated branch circuit.

    a. 2
    b. 4
    c. 5
    d. 6

24. Where compressed natural gas vehicles are repaired, the area within ____ inches of the ceiling shall be considered unclassified where adequate ventilation is provided.

    a. 18
    b. 24
    c. 30
    d. 36

25. A three-phase, 480-volt, 100-ampere noncontinuous load is located 390 feet from a panelboard. What MINIMUM size THWN aluminum conductors are required to supply the load, if the voltage drop is limited to 3 percent? (K = 21.2)

    a. 2 AWG
    b. 1 AWG
    c. 1/0 AWG
    d. 2/0 AWG

MASTER ELECTRICIAN
Practice Exam #14

The following questions are based on the 2008 edition of the National Electrical Code® and are typical of questions encountered on most Master Electricians' Exams. Select the best answer from the choices given and review your answers with the answer key included in this book.

ALLOTTED TIME: 75 minutes

Notes

Master Electrician Practice Exam #14

1. When a portable generator is used for a portable optional standby source and is not considered a separately derived system, the equipment grounding conductor shall be bonded to _____.

   a. the generator frame and the grounded conductor
   b. the system grounding electrode
   c. a grounding electrode only
   d. the grounded conductor

2. Conductors other than service-entrance conductors shall be permitted to be installed in a cable tray with service-entrance conductors provided _____.

   a. a solid fixed barrier of material compatible with the cable tray is installed to separate the systems
   b. the conductors are not in excess of 480 volts to ground
   c. the voltage between the conductors is not in excess of 600 volts
   d. the conductors have equal insulation and temperature ratings

3. When overcurrent protective devices are to be installed to protect fire pumps, they are to be sized to carry the _____ of the fire pump motor(s).

   a. starting current
   b. full-load running current
   c. stopping current
   d. locked-rotor current

4. Which of the following listed is/are prohibited from being supplied through AFCI or GFCI protective devices?

   a. 120-volt smoke alarms
   b. fire alarm systems
   c. 120-volt receptacle outlets in residential garages
   d. residential lighting outlets

5. Which of the following MUST be provided at a patient bed location used for general care in a hospital?

   a. Circuit on normal system.
   b. Circuit on emergency system.
   c. "Hospital-grade" receptacle outlets.
   d. All of these.

6. When an electrical service is required to have a grounded conductor present, what is the MINIMUM size grounded conductor permitted for an electric service using size 1000 kcmil copper ungrounded conductors?

   a. 3/0 copper
   b. 2/0 copper
   c. 1/0 copper
   d. 4/0 copper

7. Determine the MINIMUM 20-ampere, 277-volt general lighting branch circuits required for a 150,000 square foot retail department store when the actual connected lighting load is 400 kVA.

   Note: Circuit breakers of this size are not rated for continuous use.

   a. 72
   b. 82
   c. 91
   d. 102

8. When a two-gang box contains two single-pole switches (unless the box is equipped with permanently installed barriers), the voltage between the switches shall not be in excess of _____.

   a. 120 volts
   b. 277 volts
   c. 480 volts
   d. 240 volts

Notes

Notes

9. Where portions of a cable, raceway, or sleeve are known to be subjected to different temperatures and where condensation is known to be a problem, the raceway or sleeve shall _____.

    a. be filled with an approved material
    b. have an explosionproof seal
    c. be provided with a drain
    d. be installed below the level of the terminations

10. Fluorescent luminaires installed more than _____ feet above the floor level in patient care areas in hospitals, shall NOT be required to be grounded by an insulated equipment grounding conductor.

    a. 6
    b. $6\frac{1}{2}$
    c. 7
    d. $7\frac{1}{2}$

11. When rigid nonmetallic conduit (RNC) is used to enclose conductors supplying a wet-niche luminaire in a swimming pool, a size _____ insulated copper, grounding conductor shall be installed in the RNC, unless a listed low-voltage lighting system not requiring grounding is used.

    a. 12 AWG
    b. 10 AWG
    c. 8 AWG
    d. 6 AWG

12. The ampacity of phase conductors from the generator terminals to the first overcurrent device shall not be less than _____ percent of the nameplate current rating of the generator where the design of the generator does not prevent overloading.

    a. 100
    b. 115
    c. 125
    d. 150

13. When calculating the total load of an office building, and the number of general purpose receptacle outlets to be installed is not yet determined, an additional load of _____ volt-amp(s) per square foot is required for the receptacles.

    a. one
    b. two
    c. three
    d. one-half

14. An outdoor motor starter enclosure subject to temporary submersion shall be of an enclosure Type number _____.

    a. 3RX
    b. 3SX
    c. 4X
    d. 6P

15. Conductors between the controller and the diesel engine of a fire pump are required by the NEC® to be _____.

    a. 90°C rated
    b. 104°C rated
    c. stranded
    d. solid

16. The MAXIMUM allowable ampacity of a size 750 kcmil XHHW aluminum conductor when there are six current-carrying conductors in the raceway, installed in a dry location, where the ambient temperature will reach 22°C is _____.

    a. 323.40 amperes
    b. 365.40 amperes
    c. 361.92 amperes
    d. 348.00 amperes

Notes

Notes

17. Cable trays are permitted to be installed in all of the following listed locations, EXCEPT _____.

    a. basements
    b. storage rooms
    c. sealed ceiling spaces
    d. when passing through a wall

18. In compliance with the National Electric Code®, flat conductor cable (FCC) is permitted to be used for _____.

    I. general purpose branch-circuit conductors
    II. appliance branch-circuit conductors

    a. I only
    b. II only
    c. both I and II
    d. neither I nor II

19. Determine the MINIMUM size USE aluminum cable permitted for use on an underground 120/240 volt, single-phase service for a small office building that has a total load of 23,600 VA after all demand factors have been taken into consideration. Consider all conductor terminations are rated for 75°C.

    a. 1/0 AWG
    b. 2/0 AWG
    c. 1 AWG
    d. 2 AWG

20. The MINIMUM overhead clearance above the water level of a swimming pool for network-powered broadband communication systems conductors shall be _____.

    a. 14 feet
    b. 22½ feet
    c. 25 feet
    d. 27 feet

21. If an apartment complex has a total connected lighting load of 205.4 kVA, what is the demand load of this lighting in kVA?

    Given: each apartment unit will contain an electric range. (The optional method of calculation is not to be used.)

    a. 60.2 kVA
    b. 16.5 kVA
    c. 63.0 kVA
    d. 65.3 kVA

22. All of the following copper conductors are to be installed in an electrical metallic tubing (EMT) that is 40 feet long:

    - twenty-four size 10 AWG TW
    - ten size 10 AWG THW
    - fourteen size 12 AWG THHN

    Determine the MINIMUM trade size of EMT required.

    a. 2 inch
    b. 2½ inch
    c. 3 inch
    d. 3½ inch

23. A type of lighted sign the NEC® does NOT require to be listed if installed in conformance with the NEC® is a _____ sign.

    a. neon lighted
    b. portable
    c. HID lighted
    d. fluorescent lighted

24. When sizing fuses or circuit breakers for a branch-circuit serving a hermatic refrigerant motor-compressor, the device shall NOT exceed _____ of the rated load current marked on the nameplate of the equipment.

    a. 115 percent
    b. 125 percent
    c. 175 percent
    d. 225 percent

Notes

Notes

25. Given: you are to install 90 feet of multioutlet assembly in the computer lab of a school. The computers are likely to be used simultaneously. Determine the MINIMUM number of 20-ampere, 120-volt, single-phase branch-circuits required to supply the multioutlet assembly.

    a. four
    b. five
    c. six
    d. seven

MASTER ELECTRICIAN
Practice Exam #15

The following questions are based on the 2008 edition of the National Electrical Code® and are typical of questions encountered on most Master Electricians' Exams. Select the best answer from the choices given and review your answers with the answer key included in this book.

ALLOTTED TIME: 75 minutes

Notes

Master Electrician Practice Exam #15

1. The time-delay feature required for a legally required standby generator set, to avoid retransfer in case of short time re-establishment of the normal power source shall permit a setting of _____.

    a. 1 minute
    b. 5 minutes
    c. 10 minutes
    d. 15 minutes

2. In general, when conductors of different insulation ratings are installed in a common raceway and the voltage is 600 volts or less, the NEC® requires _____.

    a. all conductors shall have an insulation rating equal to at least the maximum circuit voltage applied to any conductor in the raceway
    b. the conduit allowable fill to be limited to 31 percent
    c. the conductors shall have a temperature rating of 90°C
    d. the insulation of the lower rated conductors to be identified with blue or yellow colors

3. An emergency system is required to have _____ seconds to have power available in the event of failure of the normal supply.

    a. 10
    b. 15
    c. 45
    d. 60

4. Given: a direct burial cable to be installed will have a voltage of 45 KV. The NEC® mandates the MINIMUM burial depth of the cable to be _____.

    a. 24 inches
    b. 36 inches
    c. 42 inches
    d. 48 inches

5. Elevator driving motors used with a generator field control are rated as _____ duty motors.

   a. intermittent
   b. continuous
   c. variable
   d. controlled

6. In general, on the load side of the point of grounding of a separately derived system such as a transformer, a grounded conductor is NOT permitted to be connected to _____.

   a. equipment grounding conductor
   b. normally noncurrent-carrying metal parts of equipment
   c. ground, the earth
   d. any of these

7. Given: an RV campground has a total of 150 campsites with electrical power. Twenty-five of the campsites are reserved as tent sites. How many of the sites are required to be furnished with a 20-ampere, 120-volt receptacle outlet?

   a. 105
   b. 125
   c. 150
   d. None

8. Determine the MINIMUM size type SO cord permitted to supply a 40 HP, 460-volt, three-phase, continuous-duty, AC, wound rotor motor installed in an area with an ambient temperature of 86°F.

   a. 2 AWG
   b. 4 AWG
   c. 6 AWG
   d. 8 AWG

Notes

Notes

9. Determine the full-load current of a 30-HP, three-phase, 230-volt, synchronous-motor with a 90 percent power factor.

   a. 63.0 amperes
   b. 69.3 amperes
   c. 76.23 amperes
   d. 86.62 amperes

10. A one-family dwelling unit has three ovens rated at 6, 8, and 3.5 kW, a cooktop rated at 6 kW and a broiler rated at 3.5 kW. When applying the standard method of calculations for dwellings, the demand factor, in kW, to be calculated for this kitchen equipment on the ungrounded service entrance conductors is _____.

    a. 12.2 kW
    b. 18.6 kW
    c. 27.3 kW
    d. 30.1 kW

11. When installing emergency battery pack lighting unit equipment (emergency lights), the branch-circuit supplying this equipment shall _____.

    a. be connected to the nearest receptacle outlet
    b. come from the nearest outlet of power that is compatible with the emergency lights rated voltage
    c. be fed only from an identified emergency lighting panel
    d. be on the same branch circuit serving the normal lighting in the area

12. In an industrial establishment, what is the MAXIMUM length of 200-ampere rated busway that may be tapped to a 600-ampere rated busway without providing additional overcurrent protection?

    a. 10 feet
    b. 25 feet
    c. 50 feet
    d. 75 feet

13. If a fuse label is not marked with a fuse interrupting rating, then the fuse size _____.

    a. has an interrupting rating of 100,000 amperes and is current limiting
    b. has an interrupting rating of 100,000 amperes and is not current limiting
    c. has an interrupting rating of 10,000 amperes
    d. has no interrupting rating

14. Given: a recreational vehicle park to be constructed will have a total of 40 RV sites supplied with electrical power. How many RV sites are required to be equipped with at least one 50-ampere, 125/250-volt receptacle outlet?

    a. 40
    b. 20
    c. 8
    d. 2

15. When metal wireways longer than 5 feet are mounted horizontally, they shall be secured on each end or joint and shall be secured at intervals not exceeding _____.

    a. 3 feet
    b. 4 feet
    c. 5 feet
    d. 10 feet

16. The MINIMUM burial depth required for conduit or cables installed under an airport runway, concourse or tarmac is _____.

    a. 1½ feet
    b. 2 feet
    c. 3 feet
    d. 4 feet

Notes

Notes

17. Determine the MINIMUM size overload protection required for a 480-volt, three-phase, 15 HP, continuous-duty AC motor given the following conditions:

    - design B
    - temperature rise 40° C
    - service factor – 1.12
    - actual nameplate current rating – 18 amperes

    a. 20.7 amperes
    b. 18.0 amperes
    c. 22.5 amperes
    d. 23.4 amperes

18. Given: the overload protection you have selected on the above motor is not sufficient to start the motor and trips; determine the MAXIMUM size overload protection permitted.

    a. 22.5 amperes
    b. 20.7 amperes
    c. 25.2 amperes
    d. 23.4 amperes

19. Given: a 50 kVA transformer has a 480-volt, three-phase primary and a 208/120 volt, three-phase secondary. Overcurrent protection is required on both the primary and secondary side of the transformer. Determine the MAXIMUM standard size overcurrent protection device permitted on the primary side.

    a. 125 amperes
    b. 150 amperes
    c. 175 amperes
    d. 200 amperes

20. What is the MINIMUM demand load, in VA, for the general purpose receptacles in an office building with a total of 150, 15- and 20-ampere, 120-volt receptacle outlets?

    a. 18,500 VA
    b. 10,000 VA
    c. 27,000 VA
    d. 13,500 VA

21. Determine the MAXIMUM number of size 12 AWG conductors that may be installed in a 3½ inch deep, three-gang masonry box that contains three switches.

    a. 21
    b. 23
    c. 24
    d. 27

22. A horizontal raceway entering a dust-ignition proof enclosure from one that is not, does not require a seal-off if it is _____ in length.

    a. 18 inches
    b. 5 feet
    c. 10 feet
    d. 25 inches

23. A transformer is to be installed in a retail shopping mall to serve a tenant space. A grounding electrode conductor for the separately derived system is required to be connected to the nearest available, effectively grounded _____ in the building.

    a. underground metal gas pipe
    b. structural metal member
    c. service equipment enclosure
    d. metal raceway

Notes

Notes

24. In a dairy, noncurrent-carrying metal parts of equipment MUST be grounded by _____.

    a. a copper equipment grounding conductor
    b. a driven ground rod
    c. a metal water pipe
    d. a grounded conductor

25. A dry-type transformer of less than 600 volts and NOT exceeding _____ is permitted to be installed in a hollow space of a building, such as above a lift-out acoustical ceiling, provided there is adequate ventilation.

    a. 25 kVA
    b. 37½ kVA
    c. 50 kVA
    d. 112½ kVA

SIGN ELECTRICIAN
Practice Exam #16

The following questions are based on the 2008 edition of the National Electrical Code® and are typical of questions encountered on most Sign Electricians' Exams. Select the best answer from the choices given and review your answers with the answer key included in this book.

ALLOTTED TIME: 75 minutes

Notes

Sign Electrician Practice Exam #16

1. Which of the following listed conductor insulation types listed is NOT suitable for use in a rigid PVC conduit, buried underground in direct contact with the earth?

   a. TW
   b. THHN
   c. THWN
   d. XHHW

2. As long as it has continuity to the grounding electrode system, an equipment grounding conductor may be _____.

   a. rigid metal conduit (RMC)
   b. a structural member of the building steel frame
   c. the building metallic water system
   d. a driven ground rod

3. The NEC® mandates at LEAST _____ of free conductor to be left at each junction box for the purpose of splicing conductors, or making connections to luminaires or devices.

   a. 4 inches
   b. 6 inches
   c. 8 inches
   d. 10 inches

4. Given: a 20-ampere, 120-volt, single-phase branch-circuit is to supply an electric sign; what is the MAXIMUM voltage drop the NEC® recommends on this branch-circuit?

   a. 2.4 volts
   b. 3.6 volts
   c. 5.0 volts
   d. 6.0 volts

5. Which of the following, if any, may help to reduce voltage drop in a branch-circuit?

   I. Install the conductors in a larger raceway.
   II. Increase the size of the ungrounded conductors.

   a. I only
   b. II only
   c. either I or II
   d. neither I nor II

6. When the exception of field-installed skeleton tubing (neon), all electric signs shall be _____.

   a. listed
   b. labeled
   c. provided with a 120-volt receptacle outlet
   d. provided with GFCI protection

7. Given: you are to install a rigid metal conduit (RMC) buried underground in direct contact with the earth, under a parking lot of a retail shopping mall to supply an electric sign. What is the MINIMUM burial depth of the conduit?

   a. 6 inches
   b. 12 inches
   c. 18 inches
   d. 24 inches

8. Portable or mobile electric signs shall be provided with factory installed ______.

   a. GFCI protection
   b. AFCI protection
   c. FGCI protection
   d. GACI protection

Notes

Notes

9. The required operating current for a specific luminous tube is dependent on which of the following?

   a. junction box length
   b. diameter of tubing
   c. distance from adjacent tubing
   d. branch circuit size

10. The MINIMUM size conductor permitted for wiring neon secondary circuits rated at 1,000 volts or less is _____.

   a. 14 AWG
   b. 12 AWG
   c. 16 AWG
   d. 18 AWG

11. The MAXIMUM number of size 8 AWG conductors with THW insulation permitted to be installed in a trade size 3⁄4 inch rigid schedule 40 PVC conduit, more than 2 feet in length is _____.

   a. three
   b. four
   c. five
   d. six

12. A portable electric sign shall not be placed within a pool or decorative fountain or within _____ from the inside walls of the pool or fountain.

   a. 5 feet
   b. 10 feet
   c. 15 feet
   d. 20 feet

13. Fixed or stationary electric signs installed within a decorative fountain located at a sports complex are required to be at LEAST ____ from the outside edges of the fountain.

    a. 10 feet
    b. 8 feet
    c. 6 feet
    d. 5 feet

14. Given: an electric sign is to be installed in the parking lot of a retail store and is not provided with barriers to protect it from physical damage; the NEC® requires the sign to be at LEAST _____ above the areas accessible to vehicular traffic.

    a. 10 feet
    b. 12 feet
    c. 14 feet
    d. 16 feet

15. Branch-circuits that supply electric signs or outline lighting systems shall be rated NOT to exceed _____.

    a. 15 amperes
    b. 20 amperes
    c. 30 amperes
    d. 50 amperes

16. Of the following listed, which one of the types of connector is prohibited for connection of an equipment grounding conductor to electric signs or their disconnecting means?

    a. sheet metal screws
    b. pressure connectors
    c. clamps
    d. lugs

Notes

Notes

17. Given: you are to install a service for a highway billboard electric sign consisting of three two-wire branch-circuits. The service disconnecting means shall have a rating of at LEAST _____.

    a. 15 amperes
    b. 30 amperes
    c. 60 amperes
    d. 100 amperes

18. Branch-circuit conductors installed in ballast compartments of fluorescent luminaires installed in electric signs, that are within 3 inches of a ballast, shall have a temperature rating of NOT less than _____.

    a. 105°C
    b. 90°C
    c. 75°C
    d. 60°C

19. Where branch-circuit conductors are installed in raceways, the insulated conductors LARGER than size _____ are required to be stranded conductors.

    a. 10 AWG
    b. 8 AWG
    c. 12 AWG
    d. 14 AWG

20. Switches or similar devices controlling a transformer in an electric sign must have an ampere rating of not less than _____ percent of the ampere rating of the transformer.

    a. 100
    b. 200
    c. 125
    d. 300

21. An electric sign load is considered to be continuous if the maximum current of the sign is expected to continue for ______ hour(s) or more.

    a. one-half
    b. one
    c. two
    d. three

22. When installing a sign that contains a 208-volt luminaire outside of an office building window, the NEC® requires the sign be not less than _____ feet from the window.

    a. 3
    b. 6
    c. 8
    d. 10

23. The MAXIMUM allowable ampacity of size 8 AWG THWN copper conductors installed in a 50-foot length of rigid PVC conduit is _____ when there are not more than three current-carrying conductors in the conduit.

    a. 40 amperes
    b. 50 amperes
    c. 55 amperes
    d. 60 amperes

24. Each commercial building accessible to pedestrians shall have an outside electric sign branch-circuit rated at LEAST _____ amperes that supplies no other load.

    a. 15
    b. 20
    c. 25
    d. 30

Notes

Notes

25. Luminaire lamp holders installed in sign enclosures located outdoors in wet or damp locations are required to be of the _____ type.

    a. raintight
    b. waterproof
    c. weatherproof
    d. none of these apply

SIGN ELECTRICIAN
Practice Exam #17

The following questions are based on the 2008 edition of the National Electrical Code® and are typical of questions encountered on most Sign Electricians' Exams. Select the best answer from the choices given and review your answers with the answer key included in this book.

ALLOTTED TIME: 75 minutes

Notes

Sign Electrician
Practice Exam #17

1. Given: you are to install a transformer in an attic of a retail department store and the transformer is to serve an electric sign installed outdoors in the front of the building. This installation is in compliance with the NEC® if there is an access door of at LEAST _____.

   a. 24 inches × 22½ inches
   b. 36 inches × 22½ inches
   c. 36 inches × 30 inches
   d. 48 inches × 32 inches

2. When determining the proper size branch-circuit conductors required to supply an electric sign installed with fluorescent luminaires, the calculated load shall be based on _____.

   a. the total watts of the lamps
   b. 1 VA per linear inch of the lamps
   c. the total ampere rating of the luminaires
   d. 10 VA per linear foot of the lamps

3. The NEC® requires _____ receptacle outlet(s) to be installed on an electric sign.

   a. no
   b. one
   c. two
   d. three

4. A type of electric sign the NEC® does NOT require to be listed if installed in conformance with the Code, is a _____ sign.

   a. neon lighted
   b. portable
   c. highway
   d. fluorescent lighted

5. When an electric sign is installed within a decorative fountain or within at LEAST ____ of the fountain edge, all branch-circuits supplying the sign shall have GFCI protection.

   a. 6 feet
   b. 10 feet
   c. 20 feet
   d. 50 feet

6. The maximum ampere rating of an overcurrent protection device such as a circuit breaker or a fuse protecting a branch-circuit, is directly dependent upon ____.

   a. voltage
   b. impedance
   c. current flow
   d. switch size

7. Determine the MINIMUM size listed metal box permitted by the NEC® that contains two size 12 AWG conductors and three size 8 AWG conductors.

   a. 1 inch × 1¼ inch round
   b. 3 inch × 2 inch × 1½ inch device
   c. 4 inch × 1½ inch octagon
   d. 3 inch × 2 inch × 2½ inch device

8. What is the MAXIMUM length allowed for cords supplying portable electric signs in dry locations?

   a. 10 feet
   b. 6 feet
   c. 15 feet
   d. 25 feet

Notes

Notes

9. The overcurrent protection for size 10 AWG copper conductors regardless of the temperature rating of the insulation, shall NOT exceed _____ amperes, when protecting non-motorized electric sign loads.

   a. 15
   b. 20
   c. 25
   d. 30

10. Bonding conductors provided for electric signs shall be sized at not less than size _____ copper conductors.

   a. 14 AWG
   b. 12 AWG
   c. 10 AWG
   d. 8 AWG

11. Given: you are to install a roof mounted metal sign enclosure made of sheet steel. This metal sign enclosure is required to be a MINIMUM thickness of _____.

   a. 0.202 inch
   b. 0.020 inch
   c. 0.160 inch
   d. 0.016 inch

12. In general, flexible metal conduit (FMC) shall be supported at least every _____ feet and within _____ inches of an outlet box, panelboard or disconnect switch.

   a. 4½, 12
   b. 6, 12
   c. 8, 18
   d. 10, 18

13. Schedule 40 PVC conduit shall be securely fastened within at least _____ feet of an outlet box, junction box or panelboard.

    a. 1
    b. 2
    c. 2½
    d. 3

14. If there are no junction boxes provided, what is the MAXIMUM number of 90° bends permitted in a run of Schedule 40 PVC conduit used to enclose branch-circuit conductors supplying an electric sign?

    a. two
    b. three
    c. four
    d. six

15. Given: a commercial building accessible to pedestrians will have a branch-circuit supplying an electric sign that contains neon tubing installations only. This branch-circuit shall be rated NO more than _____ amperes.

    a. 15
    b. 20
    c. 30
    d. 40

16. The NEC® requires lamps for outdoor lighting to be located below all energized conductors or other utilization equipment. An exception to this requirement would be which of the following?

    a. The lamps shall be within 6½ feet of the ground level.
    b. A lockable disconnecting means must be provided.
    c. Conductors are to be identified by orange insulation.
    d. The lamps must have an isolated grounding conductor.

Notes

Notes

17. Disregarding exceptions, when sign enclosures are supported by metal poles and the poles contain supply conductors to the sign, a handhole not less than _____ is required at the base of the pole to provide access to the supply terminations.

    a. 2 inches × 4 inches
    b. 4 inches × 4 inches
    c. 4 inches × 6 inches
    d. 6 inches × 6 inches

18. Which of the following is required, if any, for open branch circuit conductors that extend through the exterior wall of a building?

    I. The conductors are required to be in a metallic sleeve.
    II. Sleeves through the wall must slant upward from outside the building to inside the building.

    a. I only
    b. II only
    c. both I and II
    d. neither I nor II

19. What expected change in length, due to expansion, is a 200-foot run of Schedule 40 PVC conduit to have, when the PVC is installed outdoors and is exposed to an annual 90°F temperature variation from the warmest day to the coldest day?

    a. 3.65 inches
    b. 7.30 inches
    c. 10.96 inches
    d. 3.75 inches

20. The MAXIMUM number of ungrounded conductors allowed to be terminated in an individual terminal of a circuit breaker installed in a panelboard is _____.

    a. one conductor
    b. two conductors
    c. three conductors
    d. four conductors

21. Determine the MINIMUM trade size intermediate metal conduit (IMC) permitted by the NEC® to enclose three size 8 THW AWG copper branch-circuit conductors in a 95-foot conduit run.

    a. ½ inch
    b. ¾ inch
    c. 1 inch
    d. 1¼ inches

22. An electrical time clock provided for lighting branch circuits is usually connected in _____ with a single lighting circuit to be controlled.

    a. series
    b. parallel
    c. sequence
    d. tandem

23. An electrical device used to reduce voltage without changing the available power is a/an _____.

    a. rectifier
    b. amplifier
    c. transformer
    d. capacitor

24. What is the MINIMUM size copper conductor required for an overhead conductor span when supplying a pole mounted electric sign having a total load of 1,800 VA? Given:

    - 120 volt branch-circuit
    - 25 feet between supports
    - no messenger wire is provided

    a. 8 AWG
    b. 10 AWG
    c. 12 AWG
    d. 14 AWG

Notes

Notes

25. Given: a pole-mounted electric sign is to be installed at a service station having gasoline dispensing units. The sign is to be supplied by branch-circuit conductors installed in a rigid Schedule 40 PVC conduit. For the PVC to be outside of the area classified as a hazardous location, the conduit is required to be at LEAST _____ from the gasoline dispensing units.

    a. 20 feet
    b. 25 feet
    c. 30 feet
    d. 50 feet

RESIDENTIAL ELECTRICIAN
Final Exam

The following questions are based on the 2008 edition of the National Electrical Code® and are typical of questions encountered on most Residential Electricians' Exams. Select the best answer from the choices given and review your answers with the answer key included in this book. Passing score on this exam is 70 percent. The exam consists of 60 questions valued at 1.67 points each, so you must answer at least 42 questions correct for a passing score. If you do not score at least 70 percent, try again and keep studying. GOOD LUCK.

ALLOTTED TIME: 3 hours

Notes

Residential Electrician Final Exam

1. In which one of the following listed areas of a dwelling is arc-fault protection NOT required for 120-volt single-phase, 15- and 20-ampere branch circuits supplying outlets?

   a. garages
   b. dining rooms
   c. living rooms
   d. hallways

2. Where non-metallic-sheathed cable (NM) is installed in nonmetallic device boxes, the cable sheath shall extend not less than _____ inside the box and beyond any cable clamp.

   a. ⅜ inch
   b. ¾ inch
   c. ½ inch
   d. ¼ inch

3. In dwelling units, where a receptacle outlet is installed on the side or face of the basin cabinet, the outlet shall be NOT more than _____ below the countertop.

   a. 16 inches
   b. 18 inches
   c. 12 inches
   d. 24 inches

4. Given: a 240-volt, 17 kW rated electric range is to be installed in the kitchen of a one-family dwelling unit. Determine the MINIMUM size copper NM cable the NEC® requires to supply this electric range.

   a. 8 AWG
   b. 6 AWG
   c. 4 AWG
   d. 2 AWG

5. When sizing the overcurrent protection required for non-motor operated appliances such as electric ranges, water heaters, and cooktops, the overcurrent protection shall NOT exceed _____ percent of the rated current of the appliance. Where this value does not correspond to a standard overcurrent device ampere rating, the next higher standard rating shall be permitted.

   a. 115
   b. 125
   c. 150
   d. 175

6. When nails are used to fasten metal boxes to wooden studs and the nails pass through the interior of the box, the nails shall be within _____ inches of the back of the box.

   a. 1/4
   b. 1/2
   c. 3/8
   d. 3/4

7. The largest size copper conductor permitted for nonmetallic sheathed cable (NM) is _____.

   a. 4 AWG
   b. 1 AWG
   c. 2/0 AWG
   d. 2 AWG

Notes

Notes

8. For the purpose of determining conductor fill in a device box, the NEC® mandates a switch to be counted as equal to two conductors. The volume allowance for the two conductors shall be based on _____.

    a. the largest wire in the box
    b. the largest grounding conductor in the box
    c. the largest wire connected to the switch
    d. the number of clamps in the box

9. When a one-family dwelling is supplied with a single-phase, three-wire, 120/240-volt service-drop from the local utility company, the service disconnecting means shall have a rating of at LEAST _____.

    a. 60 amperes
    b. 100 amperes
    c. 150 amperes
    d. 200 amperes

10. The ampacity of type NM cable shall be in accordance with the _____ conductor temperature rating; the _____ temperature rating shall be permitted to be used for ampacity derating purposes.

    a. 60°C, 90°C
    b. 75°C, 90°C
    c. 60°C, 75°C
    d. 90°F, 75°C

11. For residential electric ranges having a rating of 8.75 kW or more, the branch-circuit rating shall be at LEAST _____.

    a. 30 amperes
    b. 40 amperes
    c. 50 amperes
    d. 60 amperes

12. Where at LEAST _____ or more NM cables containing two or more current-carrying conductors are installed,without maintaining spacing between the cables, through that same opening in wood framing that is to be fire-stopped using sealing foam, the conductors are to be derated in compliance with Table 310.15(B) (2)(a) of the NEC®.

    a. two
    b. three
    c. four
    d. five

13. Switching devices must be located at LEAST _____ feet from the inside walls of a permanently installed swimming pool unless separated from the pool by a solid fence, wall, or other permanent barrier.

    a. 5
    b. 10
    c. 15
    d. 20

14. In dwelling units, 15- or 20-ampere rated, 125-volt receptacle outlets installed within 6 feet of the outside edge of a sink in a laundry room shall _____.

    a. have a weather-proof cover
    b. have AFCI protection
    c. be a single receptacle
    d. be provided with GFCI protection

15. Range hoods shall be permitted to be cord-and-plug-connected if the receptacle outlet is accessible and supplied by an individual branch-circuit and the length of the cord does NOT exceed _____.

    a. 36 inches
    b. 24 inches
    c. 18 inches
    d. 12 inches

Notes

Notes

16. One-family dwellings with direct outdoor grade level access in front and back are required to have _____.

    I. one receptacle outlet at the back.
    II. one receptacle outlet at the front.

    a. I only
    b. II only
    c. both I and II
    d. neither I nor II

17. Underground service conductors that are not encased in concrete and buried 18 inches or more below grade shall have their location identified by a warning ribbon that is placed in the trench at LEAST ____ above the underground installation.

    a. 6 inches
    b. 12 inches
    c. 18 inches
    d. 24 inches

18. When the white-colored insulated conductor in an NM cable is used for a three-way switch loop, the conductor shall be _____.

    a. permanently reidentified
    b. a return conductor
    c. used only to supply switched outlets
    d. a grounding conductor only

19. When a 20-ampere, 120-volt receptacle outlet provided for laundry appliances is installed in the bathroom of a dwelling unit, the receptacle is required to be _____.

    a. not readily accessible
    b. a single receptacle
    c. GFCI protected
    d. supplied from the bathroom basin outlets

20. When a single branch-circuit or a multi-wire branch circuit with an equipment grounding conductor is supplying a detached garage from a house, what, if any, additional grounding electrode(s) is/are required at the garage?

    a. A driven ground rod and underground metal water pipe.
    b. A metal underground water pipe only.
    c. A metal underground water pipe and building steel, if available.
    d. No additional electrodes are required.

21. Which of the following, if any, is permitted for closing an unused circuit breaker opening in panelboards, load-centers, and switchboards?

    I. Install an unused or spare circuit breaker that is compatible with the equipment.
    II. Close the opening by installing a closure plate that fills the opening and fits securely.

    a. I only
    b. II only
    c. either I or II
    d. neither I nor II

22. In dwellings, lighting outlets shall be permitted to be controlled by occupancy sensors provided they are _________.

    a. automatic
    b. located in the hallway
    c. within 6 feet of the door(s)
    d. equipped with a manual override

23. For the purpose of sizing branch-circuits, supply fixed storage type water heaters having a capacity of 120 gallons or less, the load shall be considered a _____.

    a. noncontinuous load
    b. continuous load
    c. load having a duty-cycle of 80 percent
    d. load having a duty-cycle of 75 percent

Notes

Notes

24. Where nonmetallic sheathed cable is used with nonmetallic boxes no larger than 2¼ inches × 4 inches, the cable is not required to be secured to the box if the cable is fastened within at LEAST ____ of the box.

    a. 6 inches
    b. 8 inches
    c. 12 inches
    d. 10 inches

25. A receptacle that provides power to a pool recirculating pump motor, shall be permitted not less than _____ from the inside wall of the pool.

    a. 6 feet
    b. 5 feet
    c. 8 feet
    d. 10 feet

26. Receptacle outlets installed on a temporary service pole at a construction site are required to be _____.

    a. of the non-grounding type
    b. of the grounding type and be provided with GFCI protection
    c. rated 20 amperes
    d. of the twist-lock type and rated at least 20 amperes

27. Given: a direct buried landscape lighting circuit to be installed carries 24 volts and type UF cable is to be used as the wiring method. The cable must be buried with an earth cover of at least _____.

    a. 6 inches
    b. 12 inches
    c. 18 inches
    d. 24 inches

28. For dwelling units, the 125-volt, 15- and 20-ampere rated receptacle outlets for kitchen countertop surfaces are to be supplied with at LEAST _____ small-appliance branch-circuit(s) rated at ________.

    a. one, 20 amperes
    b. two, 20 amperes
    c. three, 15 amperes
    d. three, 20 amperes

29. When trade size ⅜ inch flexible metal conduit (FMC) is used as a fixture whip from an outlet box to a luminaire, the FMC shall not exceed _____ in length.

    a. 10 feet
    b. 4 feet
    c. 5 feet
    d. 6 feet

30. A cord-and-attachment plug-connected room air-conditioner shall not exceed _____ of the rating of the branch-circuit where no other loads such as lighting fixtures are supplied.

    a. 80 percent
    b. 75 percent
    c. 60 percent
    d. 50 percent

31. Track lighting or ceiling fans are not to be located within a MINIMUM of _____ vertically from the top of the bathtub rim or shower stall threshold.

    a. 6 feet
    b. 8 feet
    c. 4 feet
    d. 10 feet

Notes

Notes

32. When NM cable is installed through slots or holes in metal framing members, the NEC® requires which of the following for the protection of the cable?

    a. Listed bushings or grommets installed after cable installation.

    b. Listed bushings or grommets installed before cable installation.

    c. Split bushings or grommets installed after cable installation.

    d. The NEC® permits any of these installations.

33. What is the MAXIMUM length of a flexible cord that may be used for a recirculating pump motor installed on a swimming pool at a multifamily dwelling?

    a. 3 feet

    b. 4 feet

    c. 6 feet

    d. 10 feet

34. Temporary holiday decorative lighting installations for dwellings are permitted to be in use for no more than _____.

    a. 30 days

    b. 60 days

    c. 90 days

    d. 120 days

35. In kitchens of dwelling units, a receptacle outlet shall be installed at each wall countertop space that is at LEAST _____ or wider.

    a. 10 inches

    b. 12 inches

    c. 18 inches

    d. 24 inches

36. When driving a ground rod at a 45° angle and rock bottom is encountered, the electrode shall be permitted to be buried in a trench that is at LEAST _____ deep.

    a. 36 inches
    b. 30 inches
    c. 48 inches
    d. 24 inches

37. What is the MAXIMUM distance a 125-volt receptacle outlet may be installed from a hot tub installed outside at a dwelling?

    a. 20 feet
    b. 15 feet
    c. 10 feet
    d. 5 feet

38. Panelboards housing overcurrent protection devices are NOT permitted to be installed in which of the following areas of dwellings?

    a. wet locations
    b. bathrooms
    c. bedrooms
    d. detached garages

39. What is the MINIMUM number of 20-ampere rated branch-circuits required for a one-family dwelling?

    a. four
    b. two
    c. five
    d. three

Notes

Notes

40. Using the standard method of calculation for a one-family dwelling, determine the MINIMUM demand load, in VA, on the ungrounded service-entrance conductors when the residence has the following fixed appliances installed:

- water heater – 4,800 VA
- dishwasher – 1,200 VA
- garbage disposal – 1,150 VA
- trash compactor – 800 VA
- attic fan – 1,200 VA

    a. 6,863 VA
    b. 9,150 VA
    c. 8,579 VA
    d. 11,438 VA

41. In general, for size 10 AWG copper non-metallic sheathed cable (NM) the overcurrent protection shall not exceed _____.

    a. 25 amperes
    b. 20 amperes
    c. 35 amperes
    d. 30 amperes

42. Given: a one-family dwelling having a computed demand load of 175 amperes is to be provided with a 120/240-volt, single-phase, three-wire service-drop from the local utility company. What is the MINIMUM size aluminum conductors with THWN insulation permitted for use as ungrounded service-entrance conductors?

    a. 1/0 AWG
    b. 2/0 AWG
    c. 3/0 AWG
    d. 4/0 AWG

43. When a 20-ampere rated branch-circuit in a residence supplies only fixed resistance type baseboard heaters, this circuit may be loaded to a MAXIMUM value of ____.

    a. 16 amperes
    b. 20 amperes
    c. 18 amperes
    d. 14 amperes

44. If an evaporative cooler or an air-conditioner is mounted on the roof of a multifamily dwelling unit, where is the service receptacle outlet for the unit to be located?

    a. Within 75 feet of the unit.
    b. Within 50 feet of the unit and on the same level.
    c. Within 25 feet of the unit and on the same level.
    d. Not required for dwelling units.

45. What is the MINIMUM size copper equipment grounding conductor required for a 15 ampere rated branch-circuit installed in rigid metal conduit (RMC) that supplies a swimming pool circulating pump motor?

    d. 14 AWG
    b. 12 AWG
    c. 10 AWG
    d. 8 AWG

46. For a one-family dwelling, at least one receptacle outlet shall be installed in each detached garage if the garage is ____.

    a. more than 750 square feet in size
    b. less than 20 feet from the dwelling
    c. more than 20 feet from the dwelling
    d. provided with electric power

Notes

Notes

47. The service grounding electrode conductor is to be sized in accordance with the rating of the _____.

    a. main circuit breaker
    b. service-drop conductors
    c. service-entrance conductors
    d. ground rod

48. When exceptions are not taken into consideration, duplex receptacles installed in a residential garage must have _____.

    a. a weatherproof cover
    b. a double-pole circuit breaker
    c. a dedicated circuit
    d. GFCI protection

49. When wall-mounted, the required 125-volt, 15- and 20-ampere rated general purpose receptacle outlets in dwelling units, shall NOT be located more than _____ above the floor.

    a. 5½ feet
    b. 4 feet
    c. 18 inches
    d. 4½ feet

50. In general, equipment grounding conductors are to be sized accordingly based on _____.

    a. the rating of the overcurrent protective device of the circuit
    b. the size of the ungrounded conductors of the circuit
    c. full-load current of the circuit load
    d. the allowable ampacity of the ungrounded conductors after derating

51. Given: a two-gang device box will contain two size 12/2 with ground NM cables connected to a duplex-receptacle and two size 14/2 with ground NM cables connected to a single-pole switch. The two gang boxes also with contain four cable clamps. The device box is required to have a volume of at LEAST _____.

    a. 28 cubic inches
    b. 30 cubic inches
    c. 34 cubic inches
    d. 36 cubic inches

52. What is the allowable ampacity of a size 14/3 AWG Type SJO flexible cord, when only two of the conductors are considered current carrying?

    a. 15 amperes
    b. 18 amperes
    C 20 amperes
    d. 25 amperes

53. When Type NM cable is run on unfinished basement walls to switches or receptacles, the cable is required to be protected from physical damage by means of _____.

    I. electrical metallic tubing (EMT)
    II. rigid nonmetallic conduit (RNC)

    a. I only
    b. II only
    c. either I or II
    d. neither I nor II

Notes

Notes

54. Where UF cable is used as a substitute for Type NM cable for interior wiring, the conductor insulation is required to be rated at _____.

    a. 60°C
    b. 75°C
    c. 90°C
    d. 100°C

55. Non-metallic sheathed cable shall be permitted to be installed unsupported where the cable is _____.

    I. fished in existing walls of finished buildings
    II. not more than 4½ feet from the last cable support to the connection point of a luminaire

    a. I only
    b. II only
    c. neither I nor II
    d. either I or II

56. Given: a size 8 AWG NM cable is to be used to supply a 5,000-watt, 240-volt, residential cooktop. Determine the MAXIMUM standard size circuit breaker permitted for overcurrent protection for this appliance.

    a. 30 amperes
    b. 35 amperes
    c. 20 amperes
    d. 40 amperes

57. Given: you are to install several 1,000-watt, 240-volt, 6 feet long baseboard heaters in a residence. Determine the MAXIMUM number of heaters that may be supplied by a single 20-ampere, 240 volt, single-phase branch-circuit.

    a. one
    b. two
    c. three
    d. four

58. Conductors tapped from a 50 ampere branch circuit supplying electric ranges, ovens and cooktops shall have an ampacity of NOT less than _____ and shall be of sufficient size for the load to be served.

    a. 30 amperes
    b. 20 amperes
    c. 50 amperes
    d. 40 amperes

59. For maintenance, the disconnecting means for associated swimming pool _____ shall be readily accessible and within sight from the equipment to be disconnected.

    I. pump motors
    II. luminaires

    a. I only
    b. II only
    c. both I and II
    d. neither I nor II

60. When a device box is installed in a wall having a gypsum (drywall) surface, those surfaces that are broken or incomplete around the device box employing a flush-type faceplate, shall be repaired so there are no gaps or open spaces greater than _____ at the edge of the box.

    a. $1/8$ inch
    b. $1/4$ inch
    c. $3/8$ inch
    d. $1/2$ inch

Notes

JOURNEYMAN ELECTRICIAN
Final Exam

The following questions are based on the 2008 edition of the National Electrical Code® and are typical of questions encountered on most Journeyman Electricians' Exams. Select the best answer from the choices given and review your answers with the answer key included in this book. Passing score on this exam is 70 percent. The exam consists of 75 questions valued at 1.33 points each, so you must answer at least 53 questions correct for a passing score. If you do not score at least 70 percent, try again and keep studying. GOOD LUCK.

ALLOTTED TIME: 4 hours

Notes

Journeyman Electrician Final Exam

1. When determining the initial standard size time-delay fuse for branch-circuit, short-circuit, and ground-fault protection for a three-phase, continuous duty, squirrel cage, AC motor, the fuse shall have a rating of _____ of the FLC of the motor. When the value determined does not correspond to a standard ampere rating, the next higher standard rating shall be permitted.

   a. 175 percent
   b. 150 percent
   c. 225 percent
   d. 250 percent

2. In compliance with the NEC®, a conductor with three white stripes on a black background is permitted to be used as a(n) _____ conductor.

   a. ungrounded
   b. grounding
   c. grounded
   d. hot-phase

3. An installation consisting of seven non-jacketed size 12 AWG Type MC cables, each with three current-carrying conductors bundled together longer than 24 inches shall have an adjustment factor of ______.

   a. 45 percent
   b. 50 percent
   c. 60 percent
   d. 70 percent

4. Which one of the following listed wiring methods is NOT permitted to be installed in ducts or plenums fabricated to transport environmental air?

   a. flexible metallic tubing
   b. type MI cable
   c. electrical metallic tubing
   d. liquid-tight flexible metal conduit

5. When wiring gasoline dispensing pumps, the first fitting that should be installed in the raceway that emerges from below ground or concrete into the base of the gasoline dispenser is a(n) _____.

    a. automatic cut-off breakaway valve
    b. disconnect
    c. sealing fitting
    d. no fittings of any kind are permitted in this location

6. One kVA is equal to _____.

    a. 100 watts
    b. 1,000 volt-amp
    c. 100 kilo-watts
    d. 100 volt-amp

7. When determining the allowable ampacity of conductors where the conductors or cables are installed in conduits exposed to direct sunlight 4 inches above a rooftop, _____ shall be added to the expected ambient outdoor temperature to determine the applicable correction factors of Table 310.16.

    a. 60°F
    b. 40°F
    c. 30°F
    d. 25°F

8. Heavy-duty lighting track is lighting track identified for use exceeding _____.

    a. 15 amperes
    b. 600 watts
    c. 20 amperes
    d. 120 volts

Notes

Notes

9. In general, threaded entries into explosionproof equipment shall be made up with at LEAST _____ threads fully engaged.

   a. 4
   b. 5
   c. 5½
   d. 6

10. Live parts of generators operated at more than _____ volts to ground shall not be exposed to accidental contact where accessible to unqualified persons.

    a. 24
    b. 50
    c. 120
    d. 150

11. Unless listed and identified otherwise, terminations and conductors of circuits rated 100 amperes or larger, shall be rated at _____.

    a. 60°C
    b. 75°C
    c. 90°C
    d. 120°C

12. In general, all mechanical elements used to terminate a grounding electrode conductor or bonding jumper to a grounding electrode shall be accessible. Which of the following, if any, is/are an exception(s) to this rule?

    I. a connection to concrete-encased electrode.
    II. a compression connection to fire-proofed structural metal.

    a. I only
    b. II only
    c. both I and II
    d. neither I nor II

13. Regardless of the voltage of the service-entrance conductors, a clearance of not less than _____ shall be maintained between the maximum water level of a swimming pool and the service entrance conductors.

    a. 10 feet
    b. 14½ feet
    c. 19 feet
    d. 22½ feet

14. Where a galvanized eye-bolt is to be used as the point of attachment for an electrical service, the NEC® requires the eye-bolt to be installed not less than ____ feet above finished grade.

    a. 8
    b. 12
    c. 10
    d. 15

15. Under which one of the following conditions are you required to use an approved electrically conductive corrosion resistance compound on metal conduit threads?

    a. In Class I, Division 1 locations.
    b. In Class II, Division 1 locations.
    c. On field cut threads for indoor locations.
    d. On field cut threads in corrosive locations.

16. Disregarding exceptions, each patient bed location in critical care areas of hospitals shall be supplied by _____.

    I. one or more branch-circuits from the emergency system
    II. one or more branch-circuits from the normal system

    a. I only
    b. II only
    c. either I or II
    d. both I and II

Notes

Notes

17. Freestanding-type, cord-and-plug connected office partitions, shall not contain more than _____ 15-ampere, 125-volt receptacle outlets.

    a. six
    b. thirteen
    c. ten
    d. eight

18. Which one of the following listed raceways is NOT permitted for use of enclosing control wiring for electric motor-driven fire pumps?

    a. flexible metal conduit (FMC)
    b. liquidtight flexible metal conduit (LFMC)
    c. liquidtight flexible nonmetallic conduit (LFNC)
    d. intermediate metal conduit (IMC)

19. What MINIMUM size bonding conductor must bond all metal parts associated with a hydromassage bathtub, including all metal piping systems, metal parts of the electrical equipment, and pump motors?

    a. 12 AWG
    b. 10 AWG
    c. 8 AWG
    d. 6 AWG

20. When installing galvanized rigid metal conduit (RMC), where structural members readily permit fastening, the conduit is required to be fastened within _____ of a junction box or panelboard.

    a. 3 feet
    b. 5 feet
    c. 6 feet
    d. 10 feet

21. In general, flexible cords are prohibited from being installed in raceways, an exception to this rule is when ____.

    a. supplying luminaires in a public school
    b. installed in an elevator shaft of an office building
    c. installed in flexible metal conduit (FMC) above a lift-out acoustical ceiling
    d. used for protection from physical damage for industrial installations

22. Given: a 240 volt, wye-connected motor derives power from a converted single-phase source. It is desired to connect a 120/240-volt transformer for control of the motor on the load side of the converter. The control transformer must _____.

    a. be connected to the manufactured phase
    b. not be disconnected to the manufactured phase
    c. have separate overcurrent protection
    d. to be connected after start-up of the motor

23. What is the MAXIMUM size overcurrent protection device required to protect size 14 AWG conductors used for the pump motor control-circuit that is protected by the motor branch circuit protection device and extends beyond the enclosure?

    a. 15 amperes
    b. 20 amperes
    c. 45 amperes
    d. 100 amperes

24. Branch-circuits supplying more than one motor shall have an ampacity of at least _____ percent of the full-load current of the largest motor, and 100 percent of the FLC of the other motor(s) in the group.

    a. 100
    b. 115
    c. 125
    d. 150

Notes

Notes

25. When a retail furniture store has eighty continuous linear feet of display show window, the NEC® mandates at LEAST _____ receptacle outlets be provided within 18 inches of the top of the show window for the show window lighting.

 a. six
 b. seven
 c. eight
 d. nine

26. Given: you are to install fifty general-purpose receptacle outlets in an existing office having a 120/240-volt, single-phase electrical system. The receptacles are to be rated 20 amperes each, 120 volts. Determine the current, in amperes, the receptacles will add to each of the ungrounded service entrance conductors.

 a. 37.50 amperes
 b. 75.00 amperes
 c. 125 amperes
 d. 150 amperes

27. Refer to the previous question and determine the MINIMUM number of 20-ampere, 120-volt branch-circuits required for this installation.

 a. 16
 b. 12
 c. 8
 d. 4

28. When twenty size 10 AWG copper current-carrying conductors with THHN insulation are installed in a 75-foot run of trade size 1½ inch electrical metallic tubing (EMT), what is the allowable ampacity rating of each conductor?

 a. 15 amperes
 b. 20 amperes
 c. 25 amperes
 d. 30 amperes

29. Ground-fault protection shall NOT be permitted for _____.

    a. three-phase, 208/120-volt services
    b. fire pumps
    c. wound-rotor motors
    d. intermittent duty motors

30. The MINIMUM size copper conductor that may be used for grounding the secondary of an instrument transformer is _____.

    a. 10 AWG
    b. 12 AWG
    c. 8 AWG
    d. 6 AWG

31. Surface metal raceways and surface nonmetallic raceways that have removable covers are permitted to contain splices and taps; the fill is NOT to exceed _____ of the area of the raceway at the point of the splice or tap.

    a. 54 percent
    b. 60 percent
    c. 75 percent
    d. 50 percent

32. Control-circuit conductors and motor branch-circuit conductors are permitted by the NEC® to occupy the same raceway if the control-circuit conductors _____.

    a. have an applied voltage not exceeding 24 volts
    b. have a voltage to ground not to exceed 150 volts
    c. have a voltage to ground not to exceed 277 volts
    d. are functionally associated with the motor system

Notes

Notes

33. Where flexible cord is used as the wiring method to supply AC motors, the size of the conductors shall be selected in accordance with ______ of the NEC®.

    a. Table 310.16
    b. Table 310.17
    c. Table 310.18
    d. Section 400.5

34. A feeder supplying two continuous-duty, 208-volt, three-phase AC motors, one rated 10 HP and one rated 7½ HP shall have a MINIMUM ampacity of ______.

    a. 55.00 amperes
    b. 68.75 amperes
    c. 62.70 amperes
    d. 38.50 amperes

35. The reason the NEC® requires all phase conductors of the same circuit to be in the same ferrous metal raceway is to reduce _____.

    a. expense
    b. inductive heat
    c. voltage drop
    d. resistance

36. Which of the following conductor insulation types, if any, is/are acceptable for wiring in a fluorescent luminaire when the branch-circuit conductor passes within three inches of the ballast?

    I. THWs
    II. THWN

    a. I only
    b. II only
    c. both I and II
    d. neither I nor II

37. Where only Class 1 circuit conductors are in a raceway, the derating factors given in 310.15(B)(2)(a) shall apply only if the conductors _____.

   a. have an insulation rating of 60°C
   b. do not have an insulation rating of 90°C
   c. are of a size 14 AWG or less
   d. carry continuous loads in excess of 10 percent of the ampacity of each conductor

38. A general-purpose receptacle outlet is permitted to be installed within a bathtub or shower space ______.

   a. if GFCI protected
   b. if AFCI protected
   c. if provided with a watertight cover
   d. never

39. An attachment plug-and-receptacle may be permitted to serve as a motor controller if the motor is portable and has a MAXIMUM rating of ______.

   a. 1/8 HP
   b. 1/4 HP
   c. 1/3 HP
   d. 1/2 HP

40. When installing a 200-ampere rated temporary service at a construction site and the only grounding electrode available is a driven ground rod, the MINIMUM size grounding electrode conductor to ground the service to the ground rod is ______ copper.

   a. 8 AWG
   b. 4 AWG
   c. 6 AWG
   d. 3 AWG

Notes

Notes

41. Surface metal raceways are required to be secured and supported at intervals NOT exceeding _____.

    a. 4 feet
    b. 8 feet
    c. 10 feet
    d. the manufacturer's instructions

42. All spa and hot tub equipment installed in non-dwelling occupancies shall be provided with an emergency shutoff or control switch within sight of the unit(s) and located not less than _____ from the inside wall of the spa or hot tub.

    a. 6 feet
    b. 10 feet
    c. 5 feet
    d. 12 feet

43. Feeders to floating buildings are permitted to be installed in _____ where flexible connections are required.

    a. MC cable
    b. AC cable
    c. NMC cable
    d. portable power cable

44. Where a recreational vehicle park has fifty RV sites with electrical power, how many sites are required to be equipped with at least one 30-ampere, 125-volt receptacle outlet?

    a. 10
    b. 15
    c. 35
    d. 50

45. In a marina, the disconnecting means for a boat receptacle outlet must be readily accessible and not more than _____ from the receptacle outlet.

    a. 6 feet
    b. 10 feet
    c. 30 inches
    d. 50 feet

46. When installing exposed rigid metal conduit (RMC) vertically from industrial machinery, if the conduit is made up with threaded couplings, the distance between supports of the conduit shall NOT exceed _____.

    a. 3 feet
    b. 5 feet
    c. 10 feet
    d. 20 feet

47. Cablebus shall be permitted to be installed only for _____ locations.

    a. exposed
    b. commercial
    c. concealed
    d. hazardous

48. Outlet boxes intended to enclose flush devices shall have an internal depth of _____.

    a. ½ inch
    b. $^{15}/_{16}$ inch
    c. ⅞ inch
    d. 1½ inches

Notes

Notes

49. In general, with respect to the service disconnecting means, the grounding electrode conductor shall be connected to the grounded service conductor on _____.

    I. the supply side
    II. the load side

    a. I only
    b. II only
    c. either I or II
    d. both I and II

50. When a single driven ground rod is used as the grounding electrode for a temporary service and the measured resistance is greater than 25 ohms, what, if anything, does the NEC® require?

    a. Add additional electrodes until the resistance is less than 25 ohms.
    b. Add one additional electrode.
    c. Nothing is required by the NEC®.
    d. Pour salt around the electrode.

51. Fire pump supply conductors on the load side of the final disconnecting means and overcurrent device(s) are permitted to be routed through a building if ______.

    a. encased in at least 2 inches of brick or concrete
    b. installed in raceways with power conductors
    c. concealed within walls, floors, or ceilings that provide a thermal barrier of material that has at least a 15-minute fire rating
    d. installed in a listed electrical circuit protective system with a 1-hour fire rating

52. All 125-volt, single-phase, 15- and 20-ampere-rated receptacle outlets installed outdoors in a public space that is accessible to the public must ______.

    I. be GFCI protected
    II. have an enclosure that is weatherproof

    a. I only
    b. II only
    c. both I and II
    d. neither I nor II

53. Nonconductive optical fiber cable may occupy the same raceway with conductors supplying lights and power, provided the operating voltage does NOT exceed ______.

    a. 480 volts
    b. 600 volts
    c. 250 volts
    d. 120 volts

54. Shore power for boats shall be provided by single receptacles rated not less than ______.

    a. 20 amperes
    b. 30 amperes
    c. 50 amperes
    d. 60 amperes

55. When branch-circuit conductors are connecting one or more units of a data-processing system to a source of supply, the branch-circuits shall have an ampacity of at LEAST _____ of the total connected load.

    a. 80 percent
    b. 115 percent
    c. 125 percent
    d. 150 percent

Notes

Notes

56. The metal frame of a building or structure can serve as a grounding electrode under which, if any, of the following conditions?

    I. 10 feet or more of a single structural metal member is in direct contact with the earth or encased in concrete that is in direct contact with the earth.

    II. The structural metal is bonded to two driven ground rods, if the ground resistance of a single ground rod exceeds 25 ohms.

    a. I only
    b. II only
    c. either I or II
    d. neither I nor II

57. Control-circuit devices with screw-type pressure terminals used with size 14 AWG or smaller copper conductors shall be torqued to at LEAST _____ pound-inches, unless identified for a different torque value.

    a. nine
    b. seven
    c. six
    d. five

58. Straight runs of trade size 2-inch rigid metal conduit (RMC) when made up with threaded couplings shall be supported at LEAST every ______.

    a. 10 feet
    b. 14 feet
    c. 16 feet
    d. 20 feet

59. The allowable ampacity of a size 4 AWG THW copper conductor installed in a conduit with three other current-carrying conductors of the same size and insulation, in a location with an ambient temperature of 98°F is _____.

    a. 60 amperes
    b. 75 amperes
    c. 80 amperes
    d. 85 amperes

60. In general, electrical nonmetallic tubing (ENT) shall be securely fastened in place at LEAST every _____.

    a. 3 feet
    b. 5 feet
    c. 4 feet
    d. 8 feet

61. Given: a metal electrical junction box with a flat metal blank plate contains the following conductors:

    - six size 10 AWG with THWN insulation
    - ten size 12 AWG with THWN insulation
    - three size 14 AWG equipment grounding conductors

    Which of the following listed junction boxes would have the MINIMUM volume, in cubic inches, as required by the NEC®?

    a. 36 cubic inches
    b. 49 cubic inches
    c. 43 cubic inches
    d. 40 cubic inches

Notes

Notes

62. Intrinsically safe apparatus, associated apparatus, and other equipment shall be installed________, unless the equipment is a simple apparatus that does not interconnect intrinsically safe circuits.

    a. in accordance with the control drawings
    b. in the electrical equipment room
    c. on a backboard of at least ¾ inch thick plywood
    d. in a dedicated enclosure

63. Enclosures containing circuit breakers and motor controllers installed in a Class II, Division 2, hazardous location, shall be _____.

    a. dust-tight
    b. heavy-duty type
    c. rain-tight
    d. general-duty type

64. Where a 20-ampere rated branch-circuit in a residence supplies only fixed resistance-type baseboard electric heaters, this branch-circuit may be loaded to a MAXIMUM value of _____.

    a. 14 amperes
    b. 16 amperes
    c. 18 amperes
    d. 20 amperes

65. When an inverse-time circuit breaker is installed only as a disconnecting means of a single continuous-duty motor, the circuit breaker must have an ampere rating of at LEAST _____ of the full-load running current of the motor.

    a. 100 percent
    b. 125 percent
    c. 115 percent
    d. 150 percent

66. The allowable ampacity of a size 750 kcmil XHHW aluminum conductor when there are six current-carrying conductors in the raceway having a length of more than 24 inches, installed in a dry location, where the ambient temperature is 22°C is _____.

   a. 323.40 amperes
   b. 365.40 amperes
   c. 361.92 amperes
   d. 348.00 amperes

67. When calculating the service-entrance conductors for a farm service, the second largest load of the total load, shall be computed at _____.

   a. 90 percent
   b. 80 percent
   c. 65 percent
   d. 75 percent

68. Direct-buried conductors and cables emerging from the ground and up a pole shall be protected to a point above finished grade of at LEAST _____.

   a. 6 feet
   b. 8 feet
   c. 10 feet
   d. 12 feet

69. Junction boxes used in straight pulls of conductors of sizes 4 AWG and larger shall be sized so the length of the box shall not be less than _____ times the trade diameter of the largest raceway entering the box.

   a. four
   b. six
   c. eight
   d. ten

Notes

Notes

70. Which of the following listed is/are prohibited from being supplied through AFCI or GFCI protective devices?

    a. 120-volt smoke alarms
    b. Fire alarm systems
    c. 120-volt receptacle outlets in residential garages
    d. Residential lighting outlets

71. Where track lighting is installed in a continuous row, each individual section of not more than _____ feet in length shall be securely supported.

    a. 2
    b. 4
    c. 6
    d. 8

72. When conductors are adjusted in size to compensate for voltage drop, equipment grounding conductors, where required, shall be adjusted proportionally according to _____.

    a. circular mil area
    b. diameter in inches
    c. temperature rating
    d. area in square inches

73. A multiwire branch-circuit supply to a gasoline dispensing pump shall be provided with a switch that will disconnect _____.

    a. only one ungrounded supply conductor
    b. all of the ungrounded supply conductors only
    c. only the neutral (grounded) conductor
    d. the grounded conductor and all of the ungrounded supply conductors

74. Which one of the following listed wiring methods is NOT approved for use in an agricultural establishment?

    a. NMC cable
    b. UF cable
    c. NM-B cable
    d. SE cable

75. On a temporary service, where a rigid metal conduit (RMC) is used as a driven grounding electrode it shall _____.

    a. be supplemented by an additional grounding electrode if the resistance to ground is less than 25 ohms
    b. be a minimum trade size of 3⁄4 inch
    c. have an outer coating of corrosion preventing material
    d. be prohibited

Notes

Notes

Master Electrician
Final Exam

1. Given: after all demand factors have been taken into consideration for an office building, the demand load is determined to be 90,000 VA. The building has a 120/240-volt, single-phase electrical system. If the service-entrance conductors are to be installed in a buried rigid Schedule 40 PVC conduit, what MINIMUM size copper conductors with THHW insulation are required for the ungrounded service-lateral conductors?

   a. 400 kcmil
   b. 350 kcmil
   c. 300 kcmil
   d. 500 kcmil

2. Given: a commercial building is to be supplied from a transformer having a 480Y/277-volt, three-phase primary and a 208Y/120-volt, three-phase secondary. The secondary will have a balanced computed demand load of 416 amperes per phase. The transformer is required to have a MINIMUM kVA rating of _____.

   a. 100 kVA
   b. 150 kVA
   c. 86 kVA
   d. 200 kVA

3. When a motor controller enclosure is installed outdoors and is subject to be exposed to sleet, it shall have a MINIMUM rating of _____, where the controller mechanism is required to be operable when ice covered.

   a. type 3
   b. type 3S
   c. type 3R
   d. type 3SX

4. Which of the following installations would not require a disconnect within sight of the motor if the motor and motor controller are NOT within sight of each other?

   a. The disconnect is capable of being locked in the OPEN position.
   b. The disconnecting means is impractical or introduces additional hazards.
   c. Industrial installations with written safety procedures, and maintained by qualified personnel.
   d. All of these apply.

5. Which of the following statement(s) is/are true, if any, regarding fire pump motors installed at a hotel?

   I. A fire pump motor shall be permitted to be supplied by a separate service.
   II. Fire pump motor power circuits shall have automatic protection against overloads.

   a. I only
   b. II only
   c. both I and II
   d. neither I nor II

6. Health care facilities require ground fault protection _____.

   a. on all 208Y/120-volt, three-phase, four wire systems.
   b. on all essential electrical systems feeders.
   c. to be omitted from all life-safety branch-circuits.
   d. on the next level of feeders after the service disconnect.

7. What MINIMUM voltage is required after 1½ hours to the emergency lighting supplied from a storage battery if the normal source voltage of 120 volts is discontinued?

   a. 60 volts
   b. 90 volts
   c. 105 volts
   d. 120 volts

Notes

Notes

8. Determine the MAXIMUM ampere rating for an overload protective device responsive to motor current, as permitted by the NEC®, used to protect a 20-HP, 230-volt, three-phase induction type AC motor with a temperature rise of 48°C and a full-load ampere rating of 54 amperes indicated on the nameplate.

   a. 54.0 amperes
   b. 70.2 amperes
   c. 62.1 amperes
   d. 75.6 amperes

9. Hazardous (classified) atmospheres containing combustible metal dust such as aluminum or magnesium are considered to be in material Group _____ classifications.

   a. C
   b. D
   c. E
   d. F

10. Given: a 15 foot horizontal run of rigid metal conduit (RMC) is to be installed between two enclosures located in a Class II, Division 1, hazardous location. One enclosure is dust-ignitionproof; the other is not. The NEC® requires at LEAST _____ sealing fittings.

   a. none required
   b. one
   c. two
   d. three

11. Given: A junction box to be installed will contain the following:

    - three size 6 AWG, one grounded, and two ungrounded conductors
    - one size 8 AWG equipment grounding conductor
    - two internal clamps
    - one pigtail

    Determine the MINIMUM size junction box, in cubic inches, the NEC® requires.

    a. 16
    b. 18
    c. 20
    d. 23

12. Determine the MINIMUM required ampacity of the conductors supplying an elevator motor given the following related information:

    - 5 HP, 208 volts, three-phase, wound rotor
    - 15 minute rated
    - nameplate current rating is 18 amperes

    a. 15.3 amperes
    b. 16.2 amperes
    c. 18.0 amperes
    d. 22.5 amperes

13. Heating elements of cables shall be separated from the edge of outlet boxes that are to be used for mounting surface luminaires by a MINIMUM distance of _____.

    a. 12 inches
    b. 10 inches
    c. 8 inches
    d. 6 inches

Notes

Notes

14. Nonmetallic underground conduit with conductors (NUCC) larger than trade size _____ shall not be used.

    a. 2 inch
    b. 3 inch
    c. 4 inch
    d. 6 inch

15. Ceiling suspended luminaires or paddle fans located _____ or more above the maximum water level of an indoor installed spa or hot tub shall NOT require GFCI protection.

    a. 10 feet
    b. $7\frac{1}{2}$ feet
    c. 8 feet
    d. 12 feet

16. When flat conductor cable (FCC) is used for general-purpose branch circuits, the MAXIMUM rating of the circuits shall be _____ amperes.

    a. 20
    b. 30
    c. 15
    d. 10

17. Lamps located in cellulose nitrate storage vaults shall be installed in rigid luminaires of the _____ type.

    a. glass-enclosed and gasketed
    b. vapor-proof
    c. explosion proof
    d. polyvinyl-enclosed and gasketed

18. Infrared heating lamps rated at _____ watts or less shall be permitted with lampholders of the medium-base, porcelain type.

    a. 300
    b. 350
    c. 400
    d. 500

19. General purpose receptacle outlets installed in the bathroom of a hotel or motel must be _____.

    I. GFCI protected
    II. on a separate 20-ampere rated branch-circuit

    a. I only
    b. II only
    c. neither I nor II
    d. both I and II

20. What type of luminaire must be used in a totally enclosed poultry house when condensation may be present?

    a. explosion-proof
    b. watertight
    c. suitable for damp locations
    d. nonmetallic

21. When sizing overcurrent protection for fire pump motors, the device(s) shall be selected or set to carry indefinitey the ______ of the motor.

    a. locked-rotor current
    b. full-load running current
    c. full-load amperage as indicated on the motor nameplate
    d. starting current

Notes

Notes

22. Relay conductors having an applied voltage of 24 volts, and associated with a motor shall be _____.

    a. permitted to occupy the same raceway with the motor
    b. run in a separate raceway
    c. blue in color
    d. not smaller than size 12 AWG

23. When calculating the total load for a mobile home park before demand factors are taken into consideration, each individual mobile home lot shall be calculated at a MINIMUM of _____.

    a. 20,000 VA
    b. 15,000 VA
    c. 24,000 VA
    d. 16,000 VA

24. In general, on the load side of the point of grounding of a separately derived system such as a transformer, a grounded conductor is NOT permitted to be connected to _____.

    a. equipment grounding conductors
    b. normally non-current carrying metal parts of equipment
    c. ground, the earth
    d. any of these

25. Which of the following classified (hazardous) location(s), if any, does the NEC® permit flexible metal conduit (FMC) for connections to motors?

    I. Class I, Division 2
    II. Class II, Division 1

    a. I only
    b. II only
    c. Both I and II
    d. Neither I nor II

26. A type of wiring method NOT acceptable for use within agricultural buildings is _____.

    a. rigid Schedule 40 PVC conduit
    b. NMC cable
    c. jacketed Type MC cable
    d. open wiring on insulators

27. Non-metallic surface extensions shall be permitted to be run in any direction from an existing outlet, but not within _____ the floor level.

    a. 1 foot
    b. $1\frac{1}{2}$ feet
    c. 2 feet
    d. 2 inches

28. Which of the following electrical equipment is permitted to be installed in cellulose nitrate film storage vaults?

    a. portable lights
    b. receptacle outlets
    c. light switches
    d. none of these

29. Where liquidtight flexible metal conduit (LFMC) is installed in a Class I, Division 2, hazardous location, it shall be permitted to delete the bonding jumper, if the LFMC is not longer than _____ and the overcurrent protection in the circuit is limited to _____ or less and the load is not a power utilization load.

    a. 6 feet – 20 amperes
    b. 10 feet – 10 amperes
    c. 6 feet – 10 amperes
    d. 6 feet – 15 amperes

Notes

Notes

30. Which one of the following wiring methods is NOT permitted to be installed in ducts or plenums used for environmental air?

    a. rigid metal conduit (RMC)
    b. electrical metallic tubing (EMT)
    c. liquidtight flexible metal conduit (LFMC)
    d. flexible metal conduit (FMC)

31. Luminaires shall be supported independently of an outlet box where the weight exceeds _____ unless the outlet box is listed and marked for the maximum weight to be supported.

    a. 50 pounds
    b. 40 pounds
    c. 30 pounds
    d. 20 pounds

32. Color coding shall be permitted to identify intrinsically safe conductors where they are colored ____ and where no other conductors of the same color are used.

    a. light blue
    b. red
    c. yellow
    d. purple

33. The NEC® permits a building to have more than one service when _____.

    I. the load requirements of the building are at least in excess of 800 amperes
    II. the building is separated by firewalls with a 4 hour rating

    a. I only
    b. II only
    c. either I or II
    d. neither I nor II

34. Given: the kitchen of a restaurant contains the following listed cooking related equipment:

- one 14 kW range
- one 5.0 kW water heater
- one 0.75 kW mixer
- one 2.5 kW dishwasher
- one 2.0 kW booster heater
- one 2.0 kW broiler

Determine the demand load, in kW, on the service-entrance conductors for the kitchen equipment after applying the permitted demand factors.

a. 19.00
b. 26.25
c. 18.38
d. 17.06

35. Where a rooftop mounted air conditioning unit is supplied with size 8 AWG THWN copper conductors, enclosed in an electrical metallic tubing (EMT) within 3 inches of the rooftop, and exposed to direct sunlight and an ambient temperature of 100°F, the allowable ampacity of the conductors is _____.

a. 50 amperes
b. 44 amperes
c. 29 amperes
d. 25 amperes

36. What is the MINIMUM size equipment grounding conductor required for a 5 HP, three-phase, 208-volt, continuous-duty motor having 20-ampere rating overload protection and short-circuit overcurrent protection rated at 30 amperes?

a. 10 AWG
b. 14 AWG
c. 12 AWG
d. 8 AWG

Notes

Notes

37. The National Electrical Code® requires ventilation of a battery room where batteries are being charged to prevent _____.

    a. battery corrosion
    b. electrostatic charge
    c. deterioration of the building steel
    d. an accumulation of an explosive mixture

38. What is the MAXIMUM balanced demand load, in VA, permitted to be connected to a new service of a commercial building, given the following conditions?

    I. The service is 208Y/120 volts, three-phase, four-wire with a 600 ampere rated main circuit breaker.
    II. The maximum load must not exceed 80 percent of the ampere rating of the main circuit breaker.

    a. 57,600 VA
    b. 99,840 VA
    c. 172,923 VA
    d. 178,692 VA

39. Determine the MAXIMUM standard size overcurrent protection required for the primary and secondary side of a transformer, when primary and secondary overcurrent protection is to be provided, given the following related information:

    - 150 kVA rating
    - Primary – 480 volt, three-phase, three-wire
    - Secondary – 208Y/120 volt, three-phase, four-wire

    a. Primary – 500 amperes, secondary – 500 amperes
    b. Primary – 450 amperes, secondary – 600 amperes
    c. Primary – 500 amperes, secondary – 450 amperes
    d. Primary – 450 amperes, secondary – 500 amperes

Notes

40. Openings around electrical penetrations of a wall of a computer room are required to be _____.

    a. insulated
    b. airtight
    c. fire-stopped
    d. sound proof

41. The local utility company is to supply a one-family dwelling with a 120/240-volt, single-phase, messenger-supported service-drop that extends directly above the swimming pool. Where the pool has a diving board 2 feet above the water level, the service-drop conductors must have a clearance of at LEAST _____ above the diving board.

    a. $14^1/_2$ feet
    b. $16^1/_2$ feet
    c. $20^1/_2$ feet
    d. $22^1/_2$ feet

42. For a retail shopping mall, all 125-volt single-phase 15- and 20-ampere rated receptacle outlets located _____ are required to be provided with GFCI protection.

    I. in bathrooms
    II. outdoors, inaccessible to the public

    a. I only
    b. II only
    c. neither I nor II
    d. both I and II

43. Where a restaurant located in a retail shopping mall has a main circuit breaker rated at 400 amperes, the required bonding jumper to the metal water piping system is to be at LEAST size ______ copper.

    a. 6 AWG
    b. 2 AWG
    c. 3 AWG
    d. 4 AWG

Notes

44. Given: a 6,000 square foot bank building is to be constructed and the exact number of general-purpose receptacle outlets to be located in the building is unknown; the bank will also require a sign circuit. Determine the MINIMUM number of 120-volt, 20-ampere rated branch-circuits required by the NEC® for the general lighting, receptacle outlets and the sign.

   a. 12
   b. 13
   c. 14
   d. 15

45. Enclosures containing circuit breakers, switches, and motor controllers located in Class II, Division 2, hazardous locations, shall be listed as _____.

   a. gastight
   b. vapor-proof
   c. dusttight
   d. stainless steel

46. Data-processing equipment is permitted to be connected to a branch-circuit by flexible cord and attachment plug cap, if the cord does NOT exceed _____ in length.

   a. 6 feet
   b. 8 feet
   c. 10 feet
   d. 15 feet

47. Where a submersible pump is installed in a pond of a water treatment plant, the service equipment to the pump shall be no closer than _____ horizontally from the shoreline.

   a. 3 feet
   b. 5 feet
   c. 6 feet
   d. 10 feet

48. Given: a one-family dwelling to be constructed will have 4,000 square foot of livable space, a 600 square foot garage, a 400 square foot porch, a 2,000 square foot unfinished basement (adaptable for future use), three small-appliance branch circuits and one branch-circuit for the laundry room. Determine the demand load, in VA, on the ungrounded service-entrance conductors for the general lighting and receptacle loads using the standard method of calculation for a one-family dwelling.

    a. 10,350 VA
    b. 9,825 VA
    c. 7,350 VA
    d. 24,000 VA

49. Determine the MINIMUM size 75°C rated conductors permitted to supply a 25-HP, 208-volt, three-phase, Design C, fire pump motor.

    a. 4 AWG
    b. 3 AWG
    c. 2 AWG
    d. 1 AWG

50. Refer to the previous question. Determine the MAXIMUM standard size overcurrent protection device the NEC® permits to protect the fire pump motor.

    a. 200 amperes
    b. 250 amperes
    c. 350 amperes
    d. 450 amperes

51. Determine the initial permitted operational setting of an adjustable inverse time circuit breaker used for branch-circuit, short-circuit, and ground-fault protection of a 10-HP, 208-volt, three-phase, squirrel cage, Design B, continuous-duty motor.

    a. 30.8 amperes
    b. 35.7 amperes
    c. 77.0 amperes
    d. 338.8 amperes

Notes

Notes

52. Each motor controller shall be capable of _____ and _____ the motor it controls.

    a. starting, stopping
    b. starting, running
    c. stopping, disconnecting
    d. stopping, interrupting

53. When viewing the busbars from the front of a three-phase switchboard or panelboard, the "A" phase _____.

    a. is permitted to be any busbar
    b. shall be the busbar on the left
    c. shall be the busbar on the right
    d. shall be the center busbar

54. The MINIMUM size copper grounding electrode conductor for an AC service supplied with four paralleled sets of size 500 kcmil aluminum conductors is _____.

    a. 2/0 AWG
    b. 3/0 AWG
    c. 4/0 AWG
    d. 250 kcmil

55. Where one or more receptacle outlets are provided for the show window lighting in a retail store, the outlet(s) shall be installed within at LEAST _____ of the top of the show window.

    a. 12 inches
    b. 18 inches
    c. 24 inches
    d. 30 inches

56. Which of the wiring methods is permitted for use in a duct of a kitchen hood installed in a restaurant?

    a. MI cable
    b. rigid metal conduit (RMC)
    c. electrical metallic tubing (EMT)
    d. none of these

57. Given: you are to build a 200-ampere, 120/240-volt, single-phase, three-wire, overhead service for a commercial occupancy in an area where the ambient temperature reaches 110°F. Determine the MINIMUM size copper conductors required for the service-entrance conductors. Consider all terminations and conductors to be rated at 75°C.

    a. 250 kcmil
    b. 4/0 AWG
    c. 3/0 AWG
    d. 2/0 AWG

58. For a commercial or industrial establishment, what is the MINIMUM headroom required in front of a main service disconnect switch?

    a. 6 feet
    b. 6 feet, 6 inches
    c. 8 feet
    d. 10 feet

59. When using liquidtight flexible metal conduit (LFMC) to enclose service-entrance conductors and the equipment bonding jumper is routed with the LFMC, the conduit may be a MAXIMUM length of _____.

    a. 24 inches
    b. 48 inches
    c. 6 feet
    d. 10 feet

Notes

Notes

60. If two driven ground rods form the entire grounding electrode system of a temporary service, what is the MAXIMUM size copper conductor required to bond the ground rods together regardless of the size of the service?

    a. 8 AWG
    b. 6 AWG
    c. 4 AWG
    d. 2 AWG

61. Under which, if any, of the following conditions is the grounded (neutral) conductor to be considered as a current-carrying conductor?

    I. When it is only carrying the unbalanced current of a single-phase, 120/240-volt, three-wire electrical system.
    II. When it is the neutral of a three-phase, wye-connected electrical system that consists of nonlinear loads.

    a. I only
    b. II only
    c. both I and II
    d. neither I nor II

62. Luminaires using mercury vapor or metal halide lamps installed in an indoor sports area shall be ______.

    a. cord-and-plug connected
    b. stem-mounted with rigid metal conduit
    c. prohibited
    d. protected with a glass or plastic lens

63. In general, which of the following panelboards, if any, is/are required to have the grounded conductor and the grounding conductor bonded together at the panelboard?

    I. Main panelboard (used as service equipment)

    II. Sub-panelboard

    a. I only
    b. II only
    c. Both I and II
    d. Neither I nor II

64. Border lights and footlights in a theater shall be arranged so that no branch circuit supplying such equipment carries a load exceeding _____.

    a. 15 amperes
    b. 16 amperes
    c. 20 amperes
    d. 30 amperes

65. Where rigid nonmetallic conduit (RNC) is buried underground about a bulk storage facility, the conduit is required to have NOT less than _____ of cover.

    a. $1\frac{1}{2}$ feet
    b. 2 feet
    c. 3 feet
    d. 4 feet

66. Disregarding exceptions, the ampacity of the conductors from the terminals of a single-phase, 120/240-volt, 18-kW rated standby generator to the first overcurrent protection device shall have an ampere rating of NOT less than _____.

    a. 100 amperes
    b. 75 amperes
    c. 86 amperes
    d. 94 amperes

Notes

Notes

67. What is the MINIMUM height allowed for a fence enclosing an outdoor installation of 2,400 volt electrical equipment?

    a. 7 feet
    b. 8 feet
    c. 9 feet
    d. 10 feet

68. What is the MAXIMUM number of current-carrying conductors permitted in a metal wireway without derating the ampacity of the conductors?

    a. 3
    b. 20
    c. 30
    d. 42

69. Determine the MAXIMUM number of size 4/0 AWG copper conductors with THWN insulation permitted to be installed in a 4 inch × 4 inch metal wireway.

    a. 20
    b. 12
    c. 10
    d. 9

70. Which of the following conductors, if any, are permitted by the NEC® to be installed in the same raceway with the service-entrance conductors?

    a. grounding conductors
    b. sub-panel feeders
    c. branch-circuit conductors
    d. none of these

Notes

71. An aluminum grounding electrode shall be _____.

    a. installed at least 18 inches above the earth
    b. installed at least 24 inches above the earth
    c. in direct earth burial
    d. prohibited

72. Service conductors that supply a building are permitted to pass through the interior of another building ______.

    a. when the conductors are installed in rigid metal conduit (RMC)
    b. when an equipment grounding conductor is routed with the conductors
    c. if there are not more than three current-carrying conductors in the raceway
    d. never

73. Determine the MINIMUM required size 75°C conductors, installed in an area with an expected ambient temperature of 120°F, that may be used to supply a demand load of 200 amperes where served with a 208Y/120-volt, three-phase electrical system where all four conductors are considered current carrying.

    a. 250 kcmil
    b. 300 kcmil
    c. 400 kcmil
    d. 500 kcmil

74. In regard to permanently installed swimming pools, where necessary to employ flexible connections to a pool motor, ______ shall be permitted as the wiring method(s).

    I. UF cable
    II. liquidtight flexible metal conduit (LFMC)

    a. I only
    b. II only
    c. either I or II
    d. neither I nor II

Notes

75. Lighting track shall NOT be installed _____.

    I. within 3 feet horizontally of a shower or bathtub enclosure
    II. under an outdoor canopy

    a. I only
    b. II only
    c. both I and II
    d. neither I nor II

ANSWER KEY

Notes

Maintenance Electrician
Practice Exam #1 Answer Key

ANSWER	REFERENCE
1. a	90.1(B) & (C)
2. d	Art. 100
3. a	Art. 100
4. c	210.8(B)(4)
5. a	General Knowledge
6. c	Trade Knowledge
7. b	358.26
8. d	Tbl. 430.248
9. a	408.36(A)
10. c	Trade Knowledge
11. c	110.26(E)
12. d	310.3
13. b	Art. 100 200.6 (A) & (B)
14. c	240.6(A)
15. b	314.16(A)
16. d	362.30(A)
17. d	210.21(B)(1)
18. c	430.6(A)(1) Tbl. 430.248 430.22(A) FLC of 7½ HP motor = 40 amperes × 125% = 50 amperes
19. a	Tbl. 310.16
20. c	240.51(B) 240.60(D)

ANSWER	REFERENCE
21. c	Current formula
	$I = P/E \quad I = \frac{10\ \text{KW} \times 1{,}000}{240\ \text{volts}} = \frac{10{,}000\ \text{watts}}{240\ \text{volts}} = 41.6\ \text{amperes}$
22. a	Art. 100
23. a	Art. 100
24. b	Current Formula
	$I = P/E \quad I = \frac{150\ \text{watts} \times 15}{120\ \text{volts}} = \frac{2{,}250}{120} = 18.75\ \text{amperes}$
25. d	404.8(A)

Maintenance Electrician
Practice Exam #2 Answer Key

ANSWER	REFERENCE
1. d	220.18(B)
2. d	430.83(A)(2)
3. a	90.1(B)
4. b	Art.100 Trade Knowledge
5. b	422.51
6. c	Trade Knowledge
7. d	110.15
8. b	Annex C, Tbl. C.1
9. a	Tbl. 310.15(B)(2)(a)
10. b	300.20(A)
11. a	358.30(A)
12. a	General Knowledge
13. c	430.6(A)(2)
14. a	430.22(A)
15. b	110.14(C)(1)(b)(1)
16. b	240.83(D)
17. d	310.12(C) 200.6 250.119
18. a	200.7(A)(2)
19. c	3-Phase Current Formula $I = P/E \times 1.732$ $I = \frac{9\text{ KW} \times 1{,}000}{208 \times 1.732} = \frac{9{,}000\text{ watts}}{360.25} = 24.9\text{ amperes}$

ANSWER	REFERENCE
20. b	Chpt. 9, Tbl. 8 Distance Formula 1st. find allowable VD - 240 volts $\times$ 3% = 7.2 volts $D = \frac{CM \times VD}{2 \times K \times I}$ $D = \frac{16{,}510 \times 7.2}{2 \times 12.9 \times 42} = \frac{118{,}872}{1{,}083.6} = 109.7$ ft.
21. b	348.20(A)(2)c
22. d	110.26(A)(2)
23. a	210.19(A)(1)
24. c	285.3(1)
25. b	240.4(D)(5)

Residential Electrician
Practice Exam #3 Answer Key

ANSWER	REFERENCE
1. a	250.64(E)
2. b	404.14(E)
3. b	210.8(A)(2) 210.12(B)
4. d	Tbl. 402.5
5. a	210.50(C)
6. d	230.90
7. c	422.11(E)(3) 240.6(A) Current Formula I = P/E I = 4,500 VA/240 V = 18.75 amps × 150% = 28.1 amperes *NOTE: The next size circuit breaker with a rating of 30 amperes should be selected.
8. d	312.2(A)
9. a	550.10(A)
10. c	Trade Knowledge Current Formula I = P/E I = 150,000 VA/ 240 volts = 625 amperes
11. c	210.63
12. c	Tbl. 220.12 70′ × 30′ = 2,100 sq. ft. × 3 VA = 6,300 VA (house) 120 volts × 15 amperes = 1,800 VA (one circuit) 6,300 VA (house)/1,800 VA (circuit) = 3.5 = 4 circuits
13. c	Tbl. 210.21(B)(3)
14. d	230.9(A)

ANSWER	REFERENCE
15. c	Trade Knowledge Current Formula $100 \times 6 = 600$ watts total $I = P/E$ $I = 600$ watts/120 volts $= 5$ amperes
16. a	Tbl. 300.5, Note 5
17. c	314.29
18. c	314.24(B)
19. b	312.2
20. a	210.52(H)
21. a	210.52(C)(1)
22. b	Tbl. 300.5, Col. 4
23. c	334.15(C)
24. d	310.12(C) 200.6(A)-(E) 250.119
25. b	404.2(A)

Residential Electrician
Practice Exam #4 Answer Key

ANSWER	REFERENCE
1. c	210.52(C)(5)
2. d	240.24(E)
3. a	110.14(A)
4. c	240.5 (B)(2)
5. b	210.8(A)(6)
6. b	550.32(F)
7. a	680.22(A)(1)
8. d	406.3(C)
9. b	210.23(A)(1) 20 amperes × 80% = 16 amperes
10. a	314.20
11. c	230.79(C)
12. b	Tbl. 220.55, Col. B 8 KW (connected load) × 80% = 6.4 KW demand load
13. c	550.32(C)
14. a	680.23(A)(3)
15. b	Tbl. 250.66
16. c	590.3(B)
17. b	314.16(B)(3)
18. c	424.3(A)
19. b	358.30(A)

ANSWER	REFERENCE
20. a	210.11(C)(1) & (2) 220.52(A) & (B)
	2- small appliance circuits @ 1,500 VA each = 3,000 VA 1- laundry circuit @ 1,500 VA each = 1,500 VA Total = 4,500 VA
21. a	340.80
22. a	314.20
23. b	210.8(A)(7)
24. a	210.52(B)(2) 210.11(C)(2) & (3)
25. c	250.52(A)(1)

Residential Electrician
Practice Exam #5 Answer Key

ANSWER	REFERENCE
1. d	422.16(B)(2)(2)
2. b	Tbl. 310.15(B)(6)
3. c	314.27(D)
4. b	250.68(A), Ex.1
5. b	334.80 Tbl. 310.16
6. c	Tbl. 314.16(A)
7. c	220.12 Tbl. 220.12 2,600 sq. ft. × 3 VA = 7,800 VA (total lighting VA of house) 120 volts × 15 amperes = 1,800 VA (one circuit) 7,800 VA (house)/1,800 VA (one circuit) = 4.3 = 5 circuits
8. a	220.14(J)(1)
9. d	220.52(A) 1,500 VA × 2 circuits required = 3,000 VA
10. b	220.53
11. d	210.52(D)
12. c	300.4(D)
13. b	Power Formula P = I × E P = 30 amperes × 240 volts = 7,200 VA
14. d	Tbl. 310.13(A) 310.8(B) & (C)(2)
15. b	314.16(B)(4)
16. c	220.54

ANSWER	REFERENCE
17. c	210.23(B) 30 amperes × 80% = 24 amperes
18. d	406.3(A)
19. d	Tbl. 220.55 & Note 3 Use column B-2 appliances = 65% demand 5 KW + 7 KW = 12 KW (total connected load) 12 KW × 65% = 7.8 KW (demand load)
20. a	Tbl. 250.122
21. a	210.6(A)(1)
22. d	424.3(B) 210.19(A)(1)
23. c	422.35
24. b	410.116(A)(1)
25. d	680.43(B)(1)(a)

Residential Electrician
Practice Exam #6 Answer Key

ANSWER	REFERENCE
1. d	250.32(A), Ex.
2. b	230.26
3. d	430.109(B)
4. c	230.28
5. a	680.43(C)
6. c	680.22(A)(4)
7. d	314.27(B)
8. a	250.8
9. c	250.53(G)
10. d	406.8(C)
11. c	406.4(E)
12. a	210.52(C)(1)
13. b	426.4 210.19(A)(1)
14. b	Tbl. 300.5, Col. 2
15. c	210.70(A)(2)(c)
16. d	230.54(C), Ex.
17. a	300.16(A)
18. c	680.9
19. b	394.10(1)
20. d	210.8(A)(7)
21. b	334.104
22. c	Tbl. 314.16(B)

ANSWER	REFERENCE
23. b	334.30
24. a	230.31(B)
25. b	Tbl. 314.16(B) 314.16(B)(4)

$$\frac{\text{18 cu. in.(box)}}{\text{2.25 cu. in. (\#12 wire)}} = \text{8 wires (allowable fill)}$$

$$\begin{array}{l} \text{8 wires (allowable fill)} \\ \underline{-\text{2 wires (device)}} \\ \text{6 wires may be added} \end{array}$$

Size 12/2 with ground type NM = 3 wires in cable

$$\frac{\text{6 wires (allowed)}}{\text{3 wires (per cable)}} = \text{2 size 12 AWG NM cables are permitted in the box}$$

MASTER ELECTRICIAN
Final Exam

The following questions are based on the 2008 edition of the National Electrical Code® and are typical of questions encountered on most Master Electricians' Exams. Select the best answer from the choices given and review your answers with the answer key included in this book. Passing score on this exam is 75 percent. The exam consists of 75 questions valued at 1.33 points each, so you must answer at least 57 questions correct for a passing score. If you do not score at least 75 percent, try again and keep studying. GOOD LUCK.

ALLOTTED TIME: 4 hours

Journeyman Electrician
Practice Exam #7 Answer Key

ANSWER	REFERENCE
1. d	422.11(F)(3)
2. a	300.5(B)
3. d	525.32
4. b	250.53(H)
5. d	Tbl. 310.16 Tbl. 310.15(B) (2) (a) Size 1/0 AWG THW ampacity (before derating) = 150 amperes 150 amperes × 80% (adjustment factor) = 120 amperes
6. a	Art. 100
7. c	430.22(A) Tbl. 310.16 FLC of motor - 70 amperes × 125% = 87.5 amperes Size 3 AWG THW conductors with an ampacity of 100 amperes should be selected.
8. c	240.4(B)(2) & (3) 240.6(A)
9. b	Tbl. 430.250
10. d	250.104(B)
11. b	Annex C, Tbl. C.1
12. b	215.3 240 amperes × 125% = 300 amperes
13. c	350.30(A)
14. c	314.28(A)(1)
15. c	590.4(B) Tbl. 400.4
16. c	Tbl. 348.22

ANSWER	REFERENCE
17. d	680.23(B)(2)
18. c	680.10
19. a	700.12(A)
20. c	230.79(D)
21. b	210.60(B)
22. b	410.68
23. c	300.4(G)
24. c	500.5(D)
25. a	Current Formula I = P/E $I = \frac{15 \times 1{,}000}{240 \text{ volts}} = \frac{15{,}000}{240} = 62.50$ amperes

Journeyman Electrician
Practice Exam #8 Answer Key

ANSWER	REFERENCE
1. a	430.75(A)
2. d	408.7
3. d	220.14(I)
4. d	Tbl. 310.13(A) Tbl. 310.16
5. a	250.50, Ex.
6. a	517.64(A)(1), (2) & (3)
7. a	Tbl. 310.17
8. c	314.16(4)
9. d	430.6(A)(1)
10. a	220.14(F)
11. a	Annex C, Tbl. C.10(A)
12. b	110.26(E)
13. a	500.5(B)
14. d	300.6(D)
15. c	210.3
16. d	680.41
17. b	645.10
18. c	332.30
19. a	314.24(A)
20. a	300.5(D)(1)
21. B	Tbl. 250.122
22. c	230.95

ANSWER	REFERENCE
23. a	Tbl. 310.16 Tbl. 310.15(B)(2)(a) #500 KCMIL THWN copper ampacity before derating = 380 amperes 380 amperes × .67 (temp. corr.) × .7 (adj. factor) = 178.22 amperes
24. d	348.12
25. c	501.15(C)(3)

Journeyman Electrician
Practice Exam #9 Answer Key

ANSWER	REFERENCE
1. a	110.26(F)(1)(a)
2. a	210.19(A)(1), Ex. 2
3. c	Tbl. 430.250 Tbl. 430.52 430.52(C)(1), Ex.1 FLC of motor - 4.6 amperes × 300% = 13.8 amperes *NOTE: The next standard size fuse with a rating of 15 amperes should be selected.
4. b	503.5
5. d	376.56(A)
6. a	352.26
7. b	250.52(A)(5)(a)
8. d	350.12(1)
9. b	110.14(C) 310.15(A)(2) Tbl. 310.16 3 AWG THHN ampacity before derating = 110 amperes (90°C Col.) 110 amperes × .96 (temp. correction) = 105.6 amperes Size 3 AWG (60°C Col.) = 85 amperes should be selected.
10. a	230.71(A)
11. b	Tbl. 352.30
12. b	Tbl. 430.52, Note 1
13. b	392.3(B)(1)(a)
14. b	440.64
15. a	210.8(B)(2)
16. a	513.8(A)
17. d	430.24

ANSWER	REFERENCE
18. a	Tbl. 250.122
19. b	430.102(B)(2), Ex.
20. b	502.10(A)(4)
21. a	Tbl. 250.66 500 kcmil × 4 conductors = 2,000 kcmil total *NOTE: Aluminum service-entrance conductors over 1,750 kcmil require a size 3/0 AWG copper grounding electrode conductor.
22. b	Tbl. 630.11(A) Tbl. 310.16 .78 (duty cycle) × 50 amperes (primary current) = 39 amperes Tbl. 310.16 requires a size 8 AWG conductor.
23. B	647.1
24. b	517.19(B)(1) & (2)
25. d	300.13(B)

Journeyman Electrician
Practice Exam #10 Answer Key

ANSWER	REFERENCE
1. d	408.5
2. b	285.24(A)
3. c	250.52(A)(3)
4. c	300.7(A)
5. c	240.40
6. d	Chpt. 9, Tbl. 8
7. b	501.15(A)(1)
8. d	410.16(C)(1)
9. d	Tbl. 514.3(B)(1)
10. b	517.31
11. a	725.179(H)
12. b	Trade & General Knowledge 3-Phase Current Formula
13. c	680.26(B)(1)(a) & (b)
14. a	ART. 100
15. c	700.12(A)
16. c	Tbl. 210.24
17. d	430.36

Item 6:

$$VD = \frac{2KID}{CM} \qquad VD = \frac{2 \times 12.9 \times 80 \text{ amps} \times 200 \text{ ft.}}{52,620 \text{ CM}} = 7.84 \text{ volts}$$

Item 12:

$$I = \frac{\text{Power}}{E \times 1.732} \qquad I = \frac{90 \text{ KVA} \times 1,000}{208 \times 1.732} = \frac{90,000}{360.25} = 250 \text{ amperes}$$

ANSWER	REFERENCE
18. B	Tbl. 314.16(B)
19. b	550.10(G)
20. d	511.7(A)(1)
21. d	500.5(C)
22. c	422.10(A) 422.13
23. a	422.11(B)
24. d	250.53(G)
25. c	Chpt. 9, Tbl. 4 Chpt. 9, Tbl. 5

Question 18 calculation:

#12 = 2.25 cu. in. × 6 (existing wire in box) = 13.5 cubic. inches.

  27.0 cubic inches (box)
−13.5 cubic inches (existing wire in box)
  13.5 cubic inches (remaining space)

13.5 cu. in. (remaining space)/2.5 cu. in. (#10) = 5.4 or 5 wires

Question 25 calculation:

#1 THW = .1901 sq. in. × 5 = .9505 square inch
#3 THW = .1134 sq. in. × 5 = .567 square inch
Total = 1.5175 square inches

Trade size 2½ inch IMC @ 40% with a permitted fill area of 2.054 square inches should be selected.

Journeyman Electrician
Practice Exam #11 Answer Key

ANSWER	REFERENCE
1. d	110.14(C)(1)(a)(1)
2. d	250.102(C) 2,000 kcmil × 12.5% = 250 kcmil
3. d	250.4(A)(5) 250.54
4. a	430.72(C)(3)
5. c	424.38(B)(1) 424.34 424.41(H)
6. c	314.16(B)(1) Tbl. 314.16(B) 2.25 cubic inches × 2 wires = 4.50 cubic inches
7. a	334.10(2)
8. b	555.22 511.3(C)(1)(b)
9. c	Tbl. 402.5
10. c	517.18(C)
11. b	Chpt. 9, Tbl. 5 Chpt. 9, Tbl. 4

#2 THWN: 0.1158 sq. in. × 3 wires = 0.3474 square inch
#6 THWN: 0.0507 sq. in. × 4 wires = 0.2028 square inch
#10 THWN: 0.0211 sq. in. × 3 wires = 0.0633 square inch
Total = 0.6135 square inches

*NOTE: A trade size 1½ inch EMT with a permitted fill @ 40% of 0.814 square inches should be selected.

ANSWER	REFERENCE
12. d	340.112
13. a	320.80(A)
14. d	Tbl. 210.24
15. b	310.15(B)(2)(c) Tbl. 310.15(B)(2)(c)
16. c	215.3
17. d	285.3(1)
18. b	Tbl. 300.5
19. d	382.15
20. c	410.16(A) & (B)
21. a	220.14(1) 200 receptacles × 180 VA = 36,000 VA
22. d	408.36(C)
23. a	Tbl. 680.8
24. d	Tbl. 220.12 Art. 100 210.19(A)(1) 1,600 square feet × 3 VA × 125% = 6,000 VA (demand) 120 volts × 15 amperes = 1,800 VA (one circuit) 6,000 VA (load)/1,800 VA (one circuit) = 3.3 = 4 circuits
25. d	250.52(B)(2)

Master Electrician
Practice Exam #12 Answer Key

ANSWER	REFERENCE
1. c	250.30(A)(3) Tbl. 250.66 300 kcmil × 3 conductors = 900 kcmil *NOTE: Tbl. 250.66 requires a size 2/0 AWG copper grounding conductor.
2. a	300.5(K)
3. b	820.15
4. c	430.109(C)(2) 15 amperes × 80% = 12 amperes
5. d	695.7 480 volts × 85% = 408 volts
6. b	110.26(D)
7. a	240.24(B)(1)(1)
8. b	Tbl. 310.16 Tbl. 310.15(B)(2)(a) Size 250 THWN copper ampacity before derating = 255 amperes 255 amps × .82 (temp. cor.) × .8 (adj. factor) = 167.28 amperes
9. d	Tbl. 430.72(B)
10. c	250.52(B)(1) 250.104(B)
11. a	250.52(A)(4)
12. c	406.8(B) 210.8(B)(4)
13. d	Tbl.310.13(A)
14. a	ART. 100

ANSWER	REFERENCE
15. b	Tbl. 220.12
16. d	210.62
17. a	230.6(1)
18. b	430.32(A)(1)
19. d	Chpt. 9, Tbl. 8
20. c	Fig. 680.8 Tbl. 680.8
21. b	Tbl. 514.3(B)(2)
22. c	511.12
23. c	Tbl. 430.250 430.24
24. c	Tbl. 408.5
25. c	250.102(D) 250.122(C) Tbl. 250.122

Question 15 calculation:

12,000 sq. ft. × 3 VA = 36,000 VA (building)
120 volts × 15 amperes = 1,800 VA (one circuit)

$$\frac{36{,}000 \text{ VA (building)}}{1{,}800 \text{ VA (one circuit)}} = 20 \text{ circuits}$$

Question 19 calculation:

$$VD = \frac{2KID}{CM} \quad VD = \frac{2 \times 12.9 \times 90 \text{ amps} \times 225 \text{ ft.}}{52{,}620 \text{ CM}} = 9.92 \text{ volts dropped}$$

Question 23 calculation:

10 HP FLC = 30.8 amps × 125% = 38.5 amperes
$7\frac{1}{2}$ HP FLC = 24.2 amps × 100% = 24.2 amperes
Total = 62.7 amperes

Master Electrician
Practice Exam #13 Answer Key

ANSWER	REFERENCE
1. d	680.57(C)(1)
2. d	Art. 100
3. c	540.13
4. a	514.8,Ex.2
5. b	240.21(B) 240.4(B)

*NOTE: The next higher standard size overcurrent device is not permitted for feeder taps.

ANSWER	REFERENCE
6. b	314.16(B)(2), (4) & (5)

2 clamps = 1 conductor
1 device = 2 conductors
3 grounding conductors = 1 conductor
TOTAL = 4 conductors

ANSWER	REFERENCE
7. a	Art. 100
8. c	680.58
9. d	514.11(B)
10. c	517.71(A), Ex.
11. d	Tbl. 430.250 430.22(A) Tbl. 310.16

25 HP motor FLC = 74.8 amperes × 125% = 93.5 amperes
93.5 amperes/.75 (temperature correction) = 124.6 amperes
*NOTE: The wire size needs to be increased because of the elevated temperature.Size 1 AWG THWN conductors with an ampacity of 130 amperes should be selected.

ANSWER	REFERENCE
12. d	680.22(B)
13. b	517.33(A)(5)
14. b	547.9(A)(4),(5),(6) & (7)

ANSWER	REFERENCE
15. c	Tbl. 220.103
16. a	620.22(A)
17. b	551.71
18. d	675.11(A) & (C)
19. b	Current Formula Tbl. 450.3(B)

$$I = \frac{KVA \times 1{,}000}{E \times 1.732} \quad I = \frac{25{,}000\ VA}{208 \times 1.732} = \frac{25{,}000}{360.25} = 69.3\ amps \times 125\%$$
$$= 86.62\ amps$$

ANSWER	REFERENCE
20. b	250.122(F)
21. a	547.9(D)
22. c	Tbl. 430.250 430.24 Tbl. 310.16

40 HP FLC = 52 amps × 100% =	52 amperes
50 HP FLC = 65 amps × 100% =	65 amperes
60 HP FLC = 77 amps × 125% =	96 amperes
Total	213 amperes

*NOTE: Size 4/0 THWN copper with an ampacity of 230 amperes should be selected.

ANSWER	REFERENCE
23. d	430.53(A)(1)(1)
24. a	511.3(C)(2)(a)
25. c	Chpt. 9, Tbl. 8

*NOTE: 3% of 480 volts = .03 × 480 volts = 14.4 (voltage drop permitted)

$$CM = \frac{1.732 \times K \times I \times D}{VD\ permitted}$$

$$CM = \frac{1.732 \times 21.2 \times 100\ amps \times 390\ ft.}{14.4} = 99{,}446\ CM$$

Size 1/0 AWG with a circular mil area of 105,600 CM should be selected.

Master Electrician
Practice Exam #14 Answer Key

ANSWER	REFERENCE
1. b	702.10(B)
2. a	230.44, Ex.
3. d	695.5(B)
4. b	760.41(B) 760.121(B)
5. d	517.18(A) & (B)
6. b	250.24(C)(1) Tbl. 250.66
7. c	210.19(A)(1) 210.11(A) 400 kVA = 400,000 VA × 125% (continuous load) = 500,000 VA (bldg.) 277 volts × 20 amperes = 5,540 VA (one circuit) 500,000 VA. (bldg. lighting)/5540 VA (circuit) = 90.2 = 91 circuits *NOTE: Circuits need only to be installed to serve the connected load.
8. b	404.8(B)
9. a	300.7(A)
10. d	517.13(B), Ex. 2
11. c	680.23(B)(2)(b)
12. b	445.13
13. a	220.14(K)(2)
14. d	Tbl. 110.20
15. c	695.14(D)

ANSWER	REFERENCE
16. c	Tbl. 310.16 Tbl. 310.15(B)(2)(a)

#750 kcmil AL ampacity (before derating) = 435 amperes
435 amperes × 1.04 (temp. cor.) × .8 (adj. factor) = 361.92 amperes

17. c — 392.6(G)&(H)

18. c — 324.10(A)

19. c — Current Formula
Tbl. 310.16

I = P/E   I = 23/600 VA/240 volts = 98.33 amperes
*NOTE: Size 1 AWG aluminum USE cable with an ampacity of 100 amperes should be selected.

20. b — 680.8(C)
Tbl. 680.8

21. d — Tbl. 220.42

Total lighting equals 205,400 VA

1st. 3,000 VA @ 100%	=	3,000 VA
next 117,000 VA @ 35%	=	40,950 VA
remainder − 205.4 kVA − 120 kVA = 85,400 VA @ 25%	=	21,350 VA
Total	=	65,300 VA

22. a — Chpt. 9, Tbl. 5
Chpt. 9, Tbl. 4

Size 10 AWG TW	= .0243 sq. in. × 24 =	.5832 sq. in.
Size 10 AWG THW	= .0243 sq. in. × 10 =	.2430 sq. in.
Size 12 AWG THHN	= .0133 sq. in. × 14 =	.1862 sq. in.
	Total =	1.0124 sq. in.

A trade size 2 inch EMT @ 40% with a permitted fill area of 1.3242 square inches should be selected.

23. a — 600.2
600.3(A)

24. d — 440.22(A)

25. d — 220.14(H)(2)

90 ft. × 180 VA per ft. = 16,200 VA (multioutlet assembly)
20 amperes × 120 volts = 2,400 VA (one circuit)
16,200 VA (total load)/2,400 VA (one circuit) = 6.75 = 7 circuits

Master Electrician
Practice Exam #15 Answer Key

ANSWER	REFERENCE
1. d	701.11(B)(1)
2. a	300.3(C)(1)
3. a	700.12
4. c	Tbl. 300.50
5. a	620.61(B)(1)
6. d	250.30(A)
7. c	551.71
8. a	430.6(A)(1) Tbl. 430.250 430.22(A) Tbl. 400.5(A), Col.a FLC of motor = 52 amperes × 125% = 65 amperes
9. b	Tbl. 430.250 63 amperes × 1.1 (power factor) = 69.3 amperes
10. a	Tbl. 220.55 & Note 3 Use column B — 5 appliances @ 45% demand 6KW + 8KW + 3.5KW + 6KW + 3.5KW = 27KW (total connected load) 27KW × 45% = 12.15KW (demand load)
11. d	700.12(F)
12. c	368.17(B), Ex.
13. c	240.60(C)(3)
14. c	551.71 40 sites × 20% = 8 sites
15. d	376.30(A)
16. a	Tbl. 300.5

ANSWER	REFERENCE
17. a	430.6(A)(2) 430.32(A)(1) FLA of motor = 18 amperes × 115% = 20.7 amperes
18. d	430.32(C) FLA of motor = 18 amperes × 130% = 23.4 amperes
19. b	450.3(B) Tbl. 450.3(B) 240.6(A)
20. a	220.44 220.14(I) Tbl. 220.44
21. a	314.16(B)(4) Tbl. 314.16(A)
22. c	502.15(2)
23. b	250.30(A)(3) 250.30(A)(7)(2)
24. a	547.5(F)
25. c	450.13(B)

Answer 19 calculation:

$$I = \frac{KVA \times 1{,}000}{E \times 1.732} = \frac{50{,}000\ VA}{480 \times 1.732} = \frac{50{,}000\ VA}{831.36} = 60.2\ \text{amperes}$$

60.2 amperes × 250% = 150.5 amperes

Answer 20 calculation:

150 receptacles × 180 VA = 27,000 VA

1st. 10,000 VA @ 100%	= 10,000 VA
(remainder) 17,000 VA @ 50%	= 8,500 VA
TOTAL DEMAND	= 18,500 VA

Answer 21 calculation:

Masonry box: 9 size 12 AWG wires permitted per gang
−2 conductors (switch)
7 wires per box × 3 gang = 21 conductors total

Sign Electrician
Practice Exam #16 Answer Key

ANSWER	REFERENCE
1. b	300.5(B) 310.8(B) & (C)(2)
2. a	250.118(2),(3) & (4)
3. b	300.14
4. b	210.19(A)(1), FPN#4 120 volts $\times$ 3% $=$ 3.6 volts
5. b	Trade Knowledge
6. a	600.3
7. d	Tbl. 300.5
8. a	600.10(C)(2)
9. b	Trade Knowledge
10. d	600.31(B)
11. a	ANNEX C, Tbl. C.10
12. a	680.57(C)(2)
13. d	680.57(C)(1)
14. c	600.9(A)
15. b	600.5(B)(1)
16. a	250.8
17. c	230.79(D)
18. b	410.68
19. a	310.3
20. b	600.6(B)
21. d	Art. 100

ANSWER	REFERENCE
22. a	225.7(C)
23. b	Tbl. 310.16
24. b	600.5(A)
25. c	410.96

Sign Electrician
Practice Exam #17 Answer Key

ANSWER	REFERENCE
1. b	600.21(E)
2. c	220.18(B)
3. a	Trade Knowledge
4. a	600.2 600.3(A)
5. b	680.57(A) & (B)
6. c	110.9
7. c	Tbl. 314.16(B) Tbl. 314.16(A) #12 = 2.25 cu. in. × 2 = 4.5 cubic inches # 8 = 3.0 cu. in. × 3 = 9.0 cubic inches Total 13.5 cubic inches
8. c	600.10(D)(2)
9. d	240.4(D)(7)
10. a	600.7(B)(7)(a)
11. d	600.8(C)
12. a	348.30(A)
13. d	352.30(A)
14. c	352.26
15. c	600.5(B)(2)
16. b	225.25(2)
17. a	600.5(B)(3) 410.30(B)(1)
18. b	225.11 230.52

ANSWER	REFERENCE
19. b	Tbl. 352.44
	3.65 in. $\times$ 2 = 7.3 inches
20. a	110.14(A)
21. b	Annex C, Tbl. C.4
22. a	Trade Knowledge
23. c	Trade Knowledge
24. b	225.6(A)(1)
25. a	Fig. 514.3

Residential Electricians
Final Exam Answer Key

ANSWER	REFERENCE
1. a	210.12(B)
2. d	314.17(C)
3. c	210.52(D)
4. b	Tbl. 220.55 & Note 4 334.80 Tbl. 310.16 Current Formula

17 KW − 12 KW = 5 KW × 5% = 25% (increase in Col. C)
8 KW (one appliance, Col. C) × 1.25% = 10 KW demand

$$I = \frac{P}{E} \quad I = \frac{10 \text{ kw} \times 1{,}000}{240 \text{ volts}} = \frac{10{,}000}{240} = 41.66 \text{ amperes}$$

Size 6 NM cable with an ampacity of 55 amperes should be selected.

ANSWER	REFERENCE
5. c	422.11(E)(3)
6. a	314.23(B)(1)
7. d	334.104
8. c	314.16(B)(4)
9. b	230.79(C)
10. a	334.80
11. b	210.19(A)(3)
12. b	334.80
13. a	680.22(D)
14. d	210.8(A)(7)
15. a	422.16(B)(4)(2)
16. c	210.52(E)(1)

ANSWER	REFERENCE
17. b	300.5(D)(3)
18. a	200.7(C)(2)
19. c	210.8(A)(1)
20. d	250.32(A), Ex.
21. c	408.7
22. d	210.70(A)(1), Ex. #2
23. b	422.13
24. b	314.17(C), Ex.
25. a	680.22(A)(1)
26. b	590.4(D) 590.6(A)
27. a	Tbl. 300.5, Col. 5
28. b	210.11(C)(1)
29. d	410.117(C) 348.20(A)(2)c
30. a	440.62(B)
31. b	410.10(D)
32. b	300.4(B)(1)
33. a	680.21(A)(5)
34. c	590.3(B)
35. b	210.52(C)(1)
36. b	250.53(G)
37. a	680.42 680.22(A)(3)
38. b	240.24(E)
39. a	210.11(C)(1),(2) & (3)

ANSWER	REFERENCE
40. a	220.53
	4,800 VA + 1,200 VA + 1,150 VA + 800 VA + 1,200 VA = 9,150 VA TOTAL 9,150 VA × 75% (demand) = 6,862.5 VA
41. d	240.4(D)(7)
42. c	Tbl. 310.15(B)(6)
43. a	424.3(B) 210.19(A)(1)
	20 amperes/125% = 16 amperes OR 20 amperes × 80% = 16 amperes
44. c	210.63
45. b	680.21(A)(1)
46. d	210.52(G)(1)
47. c	250.66 Tbl. 250.66
48. d	210.8(A)(2)
49. a	210.52(4)
50. a	250.122(A) Tbl. 250.122
51. b	314.16(B)(1),(2),(4) & (5) Tbl. 314.16(B)

Size 14	= 2.00 cu. in. × 4 =	8.00 cu. in.
size 12	= 2.25 cu. in. × 4 =	9.00 cu. in.
equip. grnd.	= 2.25 cu. in. × 1 =	2.25 cu. in.
clamps	= 2.25 cu. in. × 1 =	2.25 cu. in.
Receptacle	= 2.25 cu. in. × 2 =	4.50 cu. in.
Switch	= 2.00 cu. in. × 2 =	4.00 cu. in.
	Total	30.00 cu. in.

*NOTE: Clamps, 1 or more, are counted as equal to the largest wire in the box 314.16(B)(2). Equipment grounding conductors, 1 or more, are counted as equal to the largest equipment grounding conductor in the box 314.16(B)(5). Devices are counted as equal to two conductors, based on the largest conductor connected to the device (314.16 (B)(4)).

ANSWER	REFERENCE
52. b	Tbl. 400.5(A), Col. B
53. c	334.15(C)
54. c	340.112
55. d	334.30(B)(1) & (2)
56. b	Tbl. 220.55 422.11(E)(3) Current Formula 240.6(A)

$I = \frac{P}{E} \quad I = \frac{5{,}000 \text{ volts}}{240 \text{ volts}} = 20.8 \text{ amperes} \times 150\% = 31.2 \text{ amperes}$

*NOTE: The next standard size circuit breaker with a rating of 35 amperes should be selected.

ANSWER	REFERENCE
57. c	424.3(B) 210.19(A)(1) Power Formula $P = I \times E$ 20 amperes × 240 volts = 4,800 VA (one circuit)

1,000 watts × 125% = 1,250 VA (demand of one heater)

$\frac{4{,}800 \text{ VA (one circuit)}}{1{,}250 \text{ VA (one heater)}} = 3.84 \text{ or } 3 \text{ heaters}$

ANSWER	REFERENCE
58. b	210.19(A)(3), Ex. #1
59. a	680.12
60. a	314.21

Journeyman Electrician
Final Exam Answer Key

ANSWER	REFERENCE
1. a	Tbl. 430.52 430.52(C)(1), Ex.1
2. c	200.7(A)(2)
3. c	310.15(B)(2), Ex.5
4. d	300.22(B)
5. c	514.9(A)
6. b	Trade Knowledge
7. c	310.15(B)(2)(c) Tbl. 310.15(B)(2)(c)
8. c	410.153
9. b	500.8(E)
10. b	445.14
11. b	110.14(C)(1)(b)(1)
12. c	250.66(A), Ex. 1 & 2
13. d	Tbl. 680.8
14. c	230.26
15. d	300.6(A)
16. d	517.18(A)
17. b	605.8(C)
18. a	695.14(E)
19. c	680.74
20. a	344.30(A)

ANSWER	REFERENCE
21. d	400.14
22. b	455.9
23. c	430.72(B)(2) Tbl. 430.72, Col. B
24. c	430.24
25. b	210.62 $\frac{\text{80 ft. (show window)}}{\text{12 ft. (per receptacle)}} = 6.7 = 7$ receptacle outlets
26. a	220.14(I) Current Formula 50 receptacles × 180 VA (each) = 9,000 VA $I = \frac{P}{E} \quad I = \frac{\text{9,000 VA}}{\text{240 volts}} = 37.5$ amperes
27. d	Power Formula $P = I \times E$ 20 amperes × 120 volts = 2,400 VA (one circuit) $\frac{\text{9,000 VA (load)}}{\text{2,400 VA (1 circuit)}} = 3.75 = 4$ circuits
28. b	Tbl. 310.16 Tbl. 310.15(B)(2)(a) Size 10 AWG THHN ampacity (before derating) = 40 amperes 40 amperes × 50% (adj. factor for 20 wires) = 20 amperes
29. b	695.6 (H)
30. b	250.178
31. c	386.56 388.56
32. d	725.48(B)(1) 300.3(C)(1)
33. d	ART. 430.6

ANSWER	REFERENCE
34. c	Tbl. 430.250 430.24 10 HP - FLC = 30.8 amps × 125% = 38.5 amperes $7\frac{1}{2}$ HP - FLC = 24.2 amps × 100% = 24.2 amperes Total = 62.7 amperes
35. b	300.20(A)
36. a	410.68 Tbl. 310.13
37. d	725.51(A)
38. d	406.8(C)
39. c	430.81(B)
40. c	250.66(A)
41. d	386.30
42. c	680.41
43. d	553.7(B)
44. c	551.71 50 sites × 70% = 35 sites
45. c	555.17(B)
46. d	344.30(B)(3)
47. a	370.3
48. b	314.24(B)
49. a	250.24(A)(5)
50. b	250.56
51. a	695.6(A) 230.6(2)
52. c	210.8(B)(4) 406.8(B)(1)
53. b	770.133

ANSWER	REFERENCE
54. b	555.19
55. c	645.5(A)
56. c	250.52(A)(2)(1) & (3) 250.56
57. b	430.9(C)
58. c	344.30(B)(2) Tbl. 344.30(B)(2)
59. a	Tbl. 310.16 Tbl. 310.15(B)(2)(a) Size 4 AWG THW ampacity = 85 amperes before derating 85 amperes × .88 (temp. cor.) × .8 (adj. factor) = 59.84 amperes
60. a	362.30(A)
61. d	314.16(B)(1) & (5) Tbl. 314.16(B) Size 10 AWG = 2.50 cu. in. × 6 wires = 15.0 cubic inches Size 12 AWG = 2.25 cu. in. × 10 wires = 22.5 cubic inches Size 14 AWG = 2.00 cu. in. × 1 wire = 2.0 cubic inches Total = 39.5 cubic inches
62. a	504.10(A)
63. a	502.115(B)
64. b	424.3(B) 210.19(A)(1) 20 amperes/125% = 16 amperes
65. c	430.110(A)
66. c	Tbl. 310.16 Tbl. 310.15(B)(2)(a) Size 750 kcmil × HHW AL ampacity (before derating) = 435 amperes 435 amperes × 1.04 (temp. cor.) × .8 (adj. factor) = 361.92 amperes

ANSWER	REFERENCE
67. d	220.103 Tbl. 220.103
68. b	300.5(D)(1)
69. c	314.28(A)(1)
70. b	760.41(B) 760.121(B)
71. b	410.154
72. a	250.122(B)
73. d	514.11(A)
74. c	547.5(A)
75. b	250.52(A)(5)(a)

Notes

Master Electrician
Final Exam Answer Key

ANSWER	**REFERENCE**
1. d	300.5(B) Tbl. 310.13(A) Tbl. 310.16 Current Formula $I = \frac{P}{E} \quad I = \frac{90{,}000 \text{ VA}}{240 \text{ volts}} = 375 \text{ amperes}$ *NOTE: Size 500 kcmil THHW conductors with an ampacity of 380 amperes should be selected.
2. b	Power Formula VA = I × E × 1.732 VA = 416 amperes × 208 volts × 1.732 = 149,866 VA $\frac{149{,}866 \text{ VA}}{1{,}000} = 149.8 \text{ kVA}$
3. b	Tbl. 110.20
4. d	430.102(B), Ex. (a) & (b)
5. a	695.3(A)(1) 695.6(D)
6. d	517.17(A)
7. c	700.12(A) 120 volts × 87.5% = 105 volts
8. b	430.6(A)(2) 430.32(C) 54 amperes × 130% = 70.2 amperes
9. c	500.6(B)(1)
10. a	502.15(2)

ANSWER	REFERENCE
11. d	314.16(B) 314.16(B)(1),(2) & (5) Tbl. 314.16(B)

2 size 6 AWG ungrounded. conductors	= 2 × 5.00 cu. in. =	10 cu. in.
1 size 6 AWG grounded conductor	= 1 × 5.00 cu. in. =	5 cu. in.
1 size 8 AWG equipt. grounding	= 1 × 3.00 cu. in. =	3 cu. in.
2 internal clamps	= 1 × 5.00 cu. in. =	5 cu. in.
1 pigtail	= 1 × 0.00 cu. in. =	0 cu. in.
	Total =	23 cu. in.

12. a — 430.22(E)
Tbl. 430.22(E)

18 amperes × 85% = 15.3 amperes
*NOTE: When sizing conductors for intermittent duty motors, the motor NAMEPLATE rating shall be used.

ANSWER	REFERENCE
13. c	424.39
14. c	354.20(B)
15. d	680.43(B)(1)(a)
16. a	324.10(B)(2)
17. a	530.51
18. a	422.48(A)
19. a	210.8(B)(1) *NOTE: Art. 210.11(C)(3) does NOT apply to hotels and motels.
20. b	547.8(C)
21. a	695.5(B)
22. a	725.48(B)(1)
23. d	550.31(1)
24. d	250.30(A)
25. a	501.10(B)(2)(2) 502.10(A)(2)
26. d	547.5(A)

ANSWER	REFERENCE
27. d	382.15(A)
28. d	530.51 & .52
29. c	501.30(B), Ex. (1) & (2)
30. c	300.22(B)
31. a	314.27(B)
32. a	504.80(C)
33. d	230.2(A),(B) & (C)
34. a	220.56 Tbl. 220.56

14.00 kw – range
5.00 kw – water heater
0.75 kw – mixer
2.50 kw – dishwasher
2.00 kw – booster heater
2.00 kw – broiler
26.25 kw – total connected load × 65% = 17.06 kw

*NOTE: However, the NEC® states the demand load shall not be less than the two largest pieces of equipment. 14 kw + 5 kw = 19 kw demand.

ANSWER	REFERENCE
35. c	310.15(B)(2)(c) Tbl. 310.15(B)(2)(c) Tbl. 310.16

Outside temp. = 100°F
Adder (3" above roof) = 40°F
Total Temp. = 140°F (for derating purposes)

Size 8 AWG ampacity (before derating) = 50 amperes
50 amperes × .58 (temperature correction) = 29 amperes

ANSWER	REFERENCE
36. a	250.122(A) Tbl. 250.122(A)
37. d	480.9(A)
38. c	Power Formula P = I × E × 1.732

P = 600 amperes × 208 volts × 1.732 = 216,154 VA
216,154 VA × 80% = 172,923 VA

ANSWER	**REFERENCE**

39. b

Three-Phase Current Formula
450.3(B)
Tbl. 450.3(B)
240.6(A)
(Primary)

$$I = \frac{VA}{E \times 1.732} \quad I = \frac{150 \times 1{,}000}{480 \times 1.732} = \frac{150{,}000}{831.36} = 180 \text{ amps} \times 250\% = 450 \text{ amperes}$$

(Secondary)

$$I = \frac{VA}{E \times 1.732} \quad I = \frac{150 \times 1{,}000}{208 \times 1.732} = \frac{150{,}000}{36.25} = 416 \text{ amps} \times 125\% = 520 \text{ amperes}$$

*NOTE: The next standard size overcurrent device with a rating of 600 amperes should be selected.

40. c

645.7
300.21

41. c

680.8
Fig. 680.8
Tbl. 680.8

22½ feet (min. ht. above water) – 2 feet (diving board) = 20½ feet

42. d

210.8(B)(1) & (4)

43. c

250.104(A)(2)
Tbl. 250.122

44. d

Art. 100
210.19(A)(1)
Tbl. 220.12
220.14(K)(2)
600.5(A)

6,000 sq. ft. × 3.5 VA = 21,000 VA × 125% = 26,250 VA (lighting)
6,000 sq. ft. × 1.0 VA = 6,000 VA (receptacles)
Total = 32,250 VA (lighting and receptacles)

120 volts × 20 amperes = 2,400 VA of one circuit

$$\frac{32{,}250 \text{ VA (load)}}{2{,}400 \text{ VA (one circuit)}} = 13.4 \text{ ckts.} = 14 \text{ lighting and receptacle circuits}$$

14 lighting and receptacle circuits + 1 sign circuit = 15 total circuits.

ANSWER	REFERENCE
45. c	502.115(B)
46. d	645.5(B)
47. b	682.11
48. a	220.12 Tbl. 220.12 220.52(A) & (B) Tbl. 220.42

4,000 sq. ft. + 2,000 sq. ft. = 6,000 sq. ft. × 3 VA = 18,000 VA
three small appliance circuits @ 1,500 VA each = 4,500 VA
one laundry circuit @ 1,500 VA = 1,500 VA
Total connected load = 24,000 VA

1st 3,000 VA @ 100% = 3,000 VA
24,000 VA − 3,000 VA = 21, 000 VA (remainder) @ 35% = 7,350 VA
Total demand load = 10,350 VA

ANSWER	REFERENCE
49. b	Tbl. 430.250 695.6(C)(2) 430.22(A) Tbl. 310.16

FLC of 25 HP, 208 volt motor = 74.8 amperes × 125% = 93.5 amperes
*NOTE: A size 3 AWG @ 75°C with an ampere rating of 100 amperes should be selected.

ANSWER	REFERENCE
50. d	695.5(B) Tbl. 430.251(B) 240.6(A)

*NOTE: The fire pump motor circuit protective device must be sized to carry indefinitely the locked-rotor current of the motor. According to Tbl. 430.251(B), the locked-rotor current of the motor is 404 amperes. The next standards size overcurrent protective device with a rating of 450 amperes should be selected.

ANSWER	REFERENCE
51. a	430.6(A)(1) Tbl. 430.250 430.52(C)(1) Tbl. 430.52

FLC of motor − 30.8 amperes × 250% = 77 amperes

ANSWER	REFERENCE
52. a	430.82(A)
53. b	408.3(E)

ANSWER	REFERENCE
54. b	Tbl. 250.66

500 kcmil × 4 conductors = 2,000 kcmil total

*NOTE: Tbl. 250.66 requires aluminum conductors of over 1,750 kcmil to have a copper grounding electrode conductor of size 3/0 AWG.

55. b	210.62
56. d	300.22(A)
57. a	Tbl. 310.16

$$\frac{\text{200 amperes (load)}}{\text{.82 (temp. correction)}} = \text{243 amperes}$$

*NOTE: Size 250 kcmil 75°C rated conductor with an ampacity of 255 amperes should be selected.

58. b	110.26(E)
59. c	230.43(15)
60. b	250.66(A)
61. b	310.15(B)(4)(a) & (c)
62. d	410.10(E)
63. a	250.24(A)(5) 408.3(C)
64. c	520.41(A)
65. b	515.8(A)
66. c	445.13 Current Formula

$$I = \frac{P}{E} \quad I = \frac{\text{18 kw} \times 1{,}000}{\text{240 volts}} = \frac{18{,}000}{240} = \text{75 amperes} \times 115\% = \text{86.25 amperes}$$

67. a	110.31
68. c	376.22(B)

ANSWER	**REFERENCE**
69. d	376.22(B) Chpt. 9, Tbl. 5
70. a	230.7, Ex.1
71. d	250.52(B)(2)
72. d	230.3
73. c	Tbl. 310.16
74. b	680.21(A)(3)
75. c	410.151(C)(2) & (9) 410.10(D) Art. 100

Calculation for 69:

4 in. x 4 in. = 16 sq. in. × 20% = 3.2 sq. in. (permitted fill)

$$\frac{\text{3.2 sq. in. (permitted fill)}}{\text{.3237 sq. in. (one wire)}} = \text{9.8 or 9 wires}$$

Calculation for 73:

$$\frac{\text{200 amperes (load)}}{\text{.75 (temp. correction)} \times \text{.8 (adjustment factor)}} = \frac{200}{.6} = \text{333 amperes}$$

*NOTE: Size 400 kcmil conductors with an ampacity of 335 amperes should be selected.

ABOUT THE AUTHOR

H. Ray Holder of San Marcos, Texas, has worked in the electrical industry for more than forty-five years as an apprentice, journeyman, master, field engineer, estimator, business manager, contractor, inspector, and instructor.

Mr. Holder is a graduate of Texas State University and holds a Bachelor of Science Degree in Occupational Education. He was awarded a lifetime teaching certificate from the Texas Education Agency in the field of Vocational Education.

He is a certified instructor of electrical trades. His classes are presented in a simplified, easy-to-understand format for electricians.

He has taught thousands of students at Austin Community College, at Austin, Texas; Odessa College, at Odessa, Texas; Howard College at San Angelo, Texas; Technical-Vocational Institute of Albuquerque, New Mexico; and in the public school systems in Ft. Worth and San Antonio, Texas. He is currently the Director of Education for Electrical Seminars, Inc. of San Marcos, Texas.

Mr. Holder is an active member of the National Fire Protection Association, International Association of Electrical Inspectors, and the International Brotherhood of Electrical Workers.